S0-AKJ-722

PRENTICE-HALL BIOLOGICAL SCIENCE SERIES

William D. McElroy and Carl P. Swanson, *Editors*

Concepts of Modern Biology Series

Animal Parasitism, CLARK P. READ

Behavioral Aspects of Ecology, 2nd ed., PETER H. KLOPFER

Concepts of Ecology, EDWARD J. KORMONDY

Critical Variables in Differentiation, BARBARA E. WRIGHT

Ecological Development in Polar Regions: A Study in Evolution, MAXWELL J. DUNBAR

Euglenoid Flagellates, GORDON F. LEEDALE

Genetic Load: Its Biological and Conceptual Aspects, BRUCE WALLACE

An Introduction to the Structure of Biological Molecules, J. M. BARRY and E. M. BARRY

Molecular Biology: Genes and the Chemical Control of Living Cells, J. M. BARRY

The Organism as an Adaptive Control System, JOHN M. REINER

Processes of Organic Evolution, G. LEDYARD STEBBINS

CONCEPTS OF MODERN BIOLOGY SERIES
William D. McElroy and Carl P. Swanson, Editors

Behavioral Aspects of Ecology

second edition

PETER H. KLOPFER

Professor of Zoology
Duke University

PRENTICE-HALL, INC., ENGLEWOOD CLIFFS, N. J.

Library of Congress Cataloging in Publication Data

KLOPFER, PETER H.
Behavioral aspects of ecology.

(Prentice-Hall biological science series. Concepts of modern biology series)

Bibliography: p.
1. Animals, Habits and behavior of. 2. Zoology—Ecology. I. Title. [DNLM: 1. Behaviorial Sciences. 2. Research. BF 76.5 K38d 1973]
QL751.K59 1973 591.5 72-6090
ISBN 0-13-073411-X

10 9 8 7 6 5 4 3 2 1

Printed in the United States of America

PRENTICE-HALL INTERNATIONAL, INC., *London*
PRENTICE-HALL OF AUSTRALIA, PTY., LTD., *Sydney*
PRENTICE-HALL OF CANADA, LTD., *Toronto*
PRENTICE-HALL OF INDIA PRIVATE LIMITED, *New Delhi*
PRENTICE-HALL OF JAPAN, INC., *Tokyo*

To Martha and Gertrud, with affection and gratitude

Contents

Series Foreword

The main body of biological literature consists of the research paper, the review article or book, the textbook, and the reference book, all of which are too often limited in scope by circumstances other than those dictated by the subject matter or the author. Unlike their usual predecessors, the books in this series, CONCEPTS OF MODERN BIOLOGY, are exceptional in their obvious freedom from such artificial limitations as are often imposed by course demands and subject restrictions.

Today the gulf of ignorance is widening, not because of a diminished capacity for learning, but because of the quantity of information being unearthed, most of which comes in small, analytical bits, undigested and unrelated. The role of the synthesizer, therefore, increases in importance, for it is he who must take the giant steps, and carry us along with him; he must go beyond his individual observations and conclusions, to assess his work and that of others in a broader context and with fresh insight. Hopefully, the CONCEPTS OF MODERN BIOLOGY SERIES provides the opportunity for decreasing the gulf of ignorance by increasing the quantity of information and quality of presentation. As editors of the Prentice-Hall Biological Science Series, we are convinced that such volumes occupy an important place in the education of the practicing and prospective teacher and investigator.

WILLIAM D. MCELROY
CARL P. SWANSON

Foreword

Three lines of investigation of animal behavior, sufficiently distinct historically and conceptually to have acquired three different names in current use, converge in this book; and it is about time they did. The separate developments of ecology, ethology, and comparative psychology constitute a fascinating nest of problems both for the history of science and for what has come to be called the sociology of science. Either kind of inquiry is, of course, much too large and exacting an enterprise to be undertaken here, but it may not be amiss to indicate a few of the extrascientific conditions that have contributed to the separateness of these fields. Among these would be linguistic barriers, culturally and nationally conditioned differences of *Weltanschauungen,* value systems, epistomologies, and so forth. For example, one might ask why the pioneer work of C. O. Whitman and Wallace Craig around the turn of the century had so little effect on comparative psychology in America that it had to wait for rediscovery in Europe many years later.

A first answer is likely to be the strong anti-instinct, or more generally antihereditarian, bias of American psychology. But why was this bias so uniquely pronounced in this country? It may well have been related to political and social convictions derived, legitimately or not, from the equalitarian beliefs of the founding fathers. Somewhat similar considerations, for example, the melioristic motive, would doubtless bear upon the early and overwhelming concentration of American comparative psychology on the problem of learning. Ethologists have frequently pointed out that to study learning without a fairly comprehensive knowledge of the animal's unlearned or species-characteristic behavior is to deprive oneself of a base line. And indeed many otherwise ingenious experiments have been largely vitiated by a failure to appreciate, for example, that a rat in a strange environment is positively thigmotactic.

Actually, relatively little of this work can properly be considered comparative psychology. Much of it was concentrated on one specific organism, the white rat, which was used simply as a cheap substitute for human subjects. With a few stalwart exceptions such as Beach, Schneirla, and Carpenter, Yerkes' early impetus, probably derived from his studies with G. H. Parker, toward a genuinely comparative psychology was diverted. Many of his earlier students, for example, became professors of education.

A list of some questions that have, at different periods, dominated the study of animal behavior by psychologists and a comparison of these with the ones raised by Dr. Klopfer in the present work will indicate the diversity of origins. Thus Romanes and Lloyd Morgan wanted to know what human mental functions or "faculties" could be found in other animals. This interest quickly focused, in Morgan, on the question: How do animals solve problems? Thorndike had a simple answer to this: by chance. Whereupon the central question became: How is a (fortuitously successful) "response" associated with a "stimulus"? These questions clearly get more and more remote from Darwin's interest in the role of behavior in speciation, from Parker's in the evolution of sense organs, and from Whitman's and the ethologist's in the characteristic action repertories of species.

But these latter questions and their specifications necessarily issue, as Dr. Klopfer indicates, in definitely psychological problems. No matter from what field the investigator starts, he cannot evade these. And it may well be that psychology's struggles with problems in areas superficially quite remote from animal behavior could prove useful to the latter field. As a matter of fact, many concepts and even theories from current ethology appear to have fairly precise isomorphs in certain theories of personality. Thus the model independently suggested by Lorenz for the quantitative study of action-specific energy is almost identical with what McDougall (1905)* called the "hydraulic analogy" he used to explain his hypothesis of inhibition by "drainage." Such independent convergence upon an hypothesis would seem, incidentally, to be a kind of evidence for the reality of something like the process envisaged, no matter how its particulars may have to be qualified or revised.

Sometimes identical terms are used. Thus the concept of hierarchical organization occurs in ethology as well as in personality theory, and is indispensable to both, as indeed to all biology. In other cases different terms have been used for what appears to be the identical or nearly identical concept. Thus the notion of releaser seems to be at least functionally equivalent to that of valence. Except for the qualification that it must be innate, the innate releasing mechanism (IRM) embodies at least one of the properties of a concept that in personality theory has had several names. Perhaps the one that comes closest to IRM is Tolman's notion of the "means-end-readiness." Contemporary American social psychology generally calls it attitude. The most extensive and coherent exploitation of the concept that I know of is that of William McDougall, who called it sentiment.

The notion of action specific energy has been little used in psychology. Owing probably to its long-standing tendency to imitate a physical style of concept formation rather than the biological, which would be more appropriate to its subject matter, the notion of energy has usually had a highly general character, such as Freud's libido or Lewin's tension. Such specification as it might show was attributed to the particular substructures through which it issued in action. The major representative of something approaching action specific energy would be, again, William McDougall with his fourteen instincts, and later, with special application to the human animal, seventeen propensities. These energies were regarded as qualitatively distinct. Of course this hypothesis is nothing like the specificity envisaged by the action specific energies of the ethologist, and further specification was

* For further information concerning references cited in the text consult the Bibliography, page 159.

attributed to the acquired differentiations, called sentiments, for objects instrumental to the discharge of the instinctive energies. In the further prosecution of this inquiry, vigorous interaction between ethologists and psychologists should be most productive.

To list all the functionally equivalent concepts and to trace out the limits of their isomorphism would take us much too far afield, but there is one correspondence that is too important to ignore. The concept of niche is clearly playing an increasingly important role in ecology and is involved in a number of the problems that Dr. Klopfer poses. The corresponding concept in psychology has had a number of names: it has been called behavioral environment, psychological environment, life-space, psychological situation, and subjective environment or subjective situation. The term subjective in these last two merely refers to the fact that the environment in question can be described only in relation to a specific organism. That is to say, a physical description of the environment will not suffice. It would seem self-evident that an animal can respond only to the environment that its sense organs, its motor equipment, and its central nervous organization, both innate and acquired, make accessible to it; and hence that any inquiry designed to describe, understand, or explain its behavior would necessarily use this environment as one of the variables that condition the behavior. But it has taken psychologists a long time to arrive at the recognition of this seemingly obvious state of affairs.

This recognition was long impeded by a conviction derived from a mistaken epistemology, namely, the conviction that we could not know the relevant, effective, or psychological environment of another animal. Actually, of course, this knowledge is subject to no epistemological disabilities that do not apply equally to our knowledge of the physical world. The latter is also dependent upon elaborate and still mysterious processes in our nervous system. It may be that some additional processes are involved in our perceptions of the world or environment of another creature, but even this is doubtful. Perceiving any object as an object involves processes which, if they were conscious, would be called reasoning. Helmholtz gave currency to the expression "unconscious inference" for such processes. And in recent years they and their kind have been called ratio-morphic processes, notably by Egon Brunswik. These processes and the bodily structures that make them possible are certainly products of evolution. They favor the survival and adaptation of the animals that possess them.

Similarly it is necessary for many higher animals to perceive the psychological environments of their conspecifics, their predators, and their prey. And just as surely the apparatus for such perception has been evolved. If it is considered that our knowledge of the environment of another is derived by analogy to our own experience of the world, then we may say that a subclass of ratio-morphic process, must also exist. This perception is just as mysterious as our perception of depth, color contrast, or objects in general *and not one bit more so.* Most psychologists, and especially behaviorists, have for a long time regarded anthropomorphism (they ought properly to say automorphism) as a methodological error rather than a mode of perception. They could hardly avoid saying such things as that the ducklings saw the hawk, or some periphrastic equivalent conveying the same idea. But they failed to realize that this was already being anthropomorphic, since "seeing" (or its behavioristic equivalent) would have no meaning if we had not experienced it ourselves. It is indeed not any more anthropomorphic to say that the ducklings see the hawk as an "air enemy." But this statement would doubtless have been rejected as a "subjective" interpretation.

We must distinguish two meanings of subjective. One has already been referred to, namely, dependence upon a relation to a specific organism. This definition has frequently been confused with the meaning, "incapable of confirmation by other observers." The term objective can hardly have any other meaning than intersubjective; and in this sense it is perfectly proper and not at all paradoxical to say that no perceptions we ever have are more objective than our perceptions of the subjective environment of another creature. Everyone who sees a given exchange between dog X and cat O would agree that X was a threat to O, and this conclusion is just as objective as the observation that X is a dog and O is a cat.

Since biology in general has thus far escaped this pitfall it may be hoped that ecology can avoid the self-defeating sterility that it has occasioned in psychology for the past half century at least. But there are already indications from the discussions in the literature of the concept of niche and of the competitive exclusion principle that a mistaken epistemological purism is beginning to rear its ugly head. The bearing of these considerations upon the questions raised by Dr. Klopfer will become, I think, abundantly evident. May the problems he has so cogently brought to our attention be vigorously attacked by ethologists and psychologists alike.

DONALD K. ADAMS
Professor of Psychology
Duke University

Preface

What is *behavioral ecology*? It is a phrase that has gained wide circulation. My interpretation was originally derived from a Yale Journal Club seminar of the 1950s, addressed by John King. He was describing his laboratory studies of the behavior of various mice, and attempting to use his findings to explain the behavior of those animals in the wild. The laboratory studies, I recall him saying, show us what is possible, that is, establish the parameters that must bound our speculations. As I believe my own work shows, this must now be regarded as only one side of the coin. An animal's capacities vary with its situation—the context of the tests to which it is subjected. Thus, not only may lab studies reveal hidden capacities to the field worker, but also the latter's observation may provide revelations to the psychologists. Behavioral ecology, in short, is best not regarded as a field so much as an admission that what we see varies with the angle of view.

Behavioral ecology implies a readiness to accept the discomfort that changing postures may impose.

In the preface to the first edition I wrote the following:

> Zoology has not been renowned for a paucity of methods or viewpoints. Some workers have approached ecological problems from a theoretical, mathematical standpoint; others have been decidedly pragmatic and empirical. A few have even proven able to reconcile these often divergent approaches within the scope of some particularly masterful study (for example, MacArthur, 1958). Yet, for all the diversity in the approaches to the major problems of ecology, there has been a striking neglect of psychological factors that control or regulate the behavior of animals. This bald assertion is not intended to minimize the frequency, importance, or value of specific behavioral studies which seek to explain habitat selection, food preferences, or the like. What is lacking is a more general account of the relation between the principles and facts studied by psychologists and those of interest to ecologists. In this short book, I hope to summarize what I consider to be the major problems of ecology and to suggest how the application of psychological viewpoints can contribute to our understanding of them. The approach is frankly speculative, for the present need seems to be for a suggestion of the nature of the rapprochement of ecology and psychology rather than for an exhaustive review of the areas of overlap.
>
> What, then, are the ecologist's fundamental problems? These have traditionally dealt with the manner in which a finite amount of space and energy is distributed among species, as well as with a temporal dimension of this distribution. We may restate these problems colloquially:
>
> 1. Why don't predators overeat their prey?
> 2. How are space and food shared?
> 3. Why are there so many species?
> 4. How do species remain distinct?
> 5. How are communities organized?

Since I wrote these lines, the questions I listed have lost some of their cogency. Nor does the application of psychology's insights to biological problems seem as novel or daring as it once did. Nevertheless, I am persuaded that undergraduates and beginning graduates will find some profit in this volume's eclectic approach to behavior and ecology. It is unlikely that any will be misled into believing that I have provided either final answers or a definitive review to the questions I posed. I do hope, however, they will be led to refine those questions and seek new vantages from which to view them.

I acknowledge with gratitude the secretarial assistance and technical help of Catherine Dewey. Sara Horowitz, Jane Spiegel, Lee McGeorge, and Robert Fudge have also been of great help. My research has been supported by a Research Scientist Award and Grant MH04453 from the NIMH and HD02319 from the NICHD. The colleagues and students who have stimulated much of the work are too numerous to list individually, but they will know I am grateful, especially J. P. Hailman who contributed a portion of one chapter.

The acknowledgment to the first edition remains appropriate:

> I wish to acknowledge with special gratitude the encouragement of R. H. MacArthur, whose ideas inspired this work. Indeed, it is often difficult for me to decide where his ideas leave off and my own begin. My teacher, Professor G. Evelyn Hutchinson, and Dr. Margaret Mead have also contributed much to the development of my ideas, and to them, also, thanks are due. I am grateful to the Chairman of the Zoology Department at Duke University, Professor E. C. Horn, for having lightened my teaching load sufficiently to allow time for these speculations, and to Miss Julie Hilger for invaluable editorial assistance. J. P. Hailman and J. J. Hatch critically read the manuscript and proofs, for which they have my thanks. Financial support for some of the studies reported herein, as well as secretarial assistance, has been provided by grants from The National Institutes of Mental Health and the Mae Smith Fund. . . .

Peter H. Klopfer

Behavioral Aspects of Ecology

ONE

Why Don't Predators Overeat Their Prey?

Predators, except for man, do not generally eradicate their prey. Indeed, if this were not so, complete extinction of both the predator and its prey would soon ensue, the elimination of the prey species leaving the predator to die by starvation. Extreme oscillations in the numbers of prey and predator that result in mutual extinction have, of course, been observed in a number of laboratory experiments, such as those on prey–predator relationships by Gause (1934) or Utida (1957). Periodically such oscillations may occur in nature, usually within circumscribed regions of the globe, particularly the northern latitudes. The oscillations in the numbers of snowy owls (*Nyctea* species) and their rodent and avian preys provide an example, as may the oscillations that exist in the numbers of arctic rabbits (*Oryctolagus* sp.) and arctic foxes (*Alopex* sp.) or, for that matter, in rabbits and their vegetational "prey" (cf. Odum, 1959). Some workers (Keith,

1963) have questioned whether these oscillations are in fact due to predator–prey interactions. In general, however, there is little question but that a balance between the numbers of prey and predator is maintained. The extreme oscillations of the northern latitudes are striking precisely because they are unusual.

Needless to say, we shall count many more oscillations if we include those produced by human intervention. The history of the use of DDT and other insecticides, which, by destroying natural predators ultimately result in massive blooms of the very creatures they were designed to kill, is a case in point. Mendelssohn (personal communication) and his collaborators at the University of Tel Aviv have recorded another intriguing example that stems from an effort at rabies control by Israeli health officials. Poisoned bait was widely distributed to eliminate the wild jackals (*Canis* sp.) on the outskirts of Tel Aviv. This led as well to the demise of the mongoose (*Herpestes* sp.). Coincident with the drop in the number of jackals and mongooses, the Tel Aviv hospital records show a steady rise in the number of snakebite victims. The incidence of such bites had been falling steadily up to the time the rabies campaign began.

By and large, however, a balance is maintained between prey and predator through processes that will be discussed in the following pages. The very fact that the extreme oscillations of the northern latitudes are so striking attests to the more usual incidence of homeostatic mechanisms that protect prey from exhaustion. The questions with which we wish to deal concern the nature of this balance, or equilibrium. Why, in short, don't predators overeat their prey?

INTRODUCTION TO THE PROBLEM

The question actually has two aspects: (1) Why doesn't evolution produce overly efficient predator genotypes, and (2) why don't individual predators overeat and then starve?

We may make two sorts of explanations. On the one hand, it is possible to conceive statistical explanations that account for the stabilization of prey and predator. Alternatively, it is possible to conceive strictly behavioral or individual adaptations that prevent any one prey from being eaten by any one predator. Our explanations may thus be based on the behavior of the individual particle or on the aggregate of particles. These distinctions merely reflect approaches at two different levels. Into the category of statistical explanations fall such studies as those by Nicholson and Bailey (1935), or Volterra (1931), and others who interpret the equilibrium of natural populations as being due to a density-dependent mortality. That is, as the abundance of the prey increases, the proportion of the prey taken by the predator also increases. As the prey becomes scarcer, the proportion taken falls off. The predator populations may be maintained at a moderately constant level by their having alternative prey available to which they may switch when the most preferred prey becomes too scarce. Indeed, MacArthur (1955) has given a mathematical demonstration of how the stability of the community may be in large part a function of the number of alternative prey available to a given predator. Where the food chains are of the linear variety, maximum instability is attained. Where true food webs exist with many lines from lower energy levels leading to a single higher level, that is, where there are many alternative forms of energy available to the predator, a maximum degree of stability is attained and the magnitude of the periodic oscillations that may occur is damped.

An experimental test of this model was devised by Paine (1966, 1969), who removed the chief predator, a starfish (*Asterias* sp.), from a marine community. Among the many other changes that followed the removal of the predator was a striking decrease in species diversity. This important experiment might well be repeated in a variety of situations.

Additional examples of "statistical" explanations are those of Huffaker et al. (1963), who finds prey–predator systems stable so long as spatial heterogeneity exists. Prey presumably find better hiding places in more varied environments so that as the prey become rarer, the level of predation becomes less intense. Pimentel (1961) devised a model that also considers changes in gene frequency. As numbers of a prey or predator change, so might the selective pressures operating on them. Traits that are advantageous under one set of densities could be disadvantageous under another. Changes in the frequency of particular genes could result, and this, in turn, could provide a feedback control of density and diversity.

Into the particle-behavior category fall studies of behavioral adaptations that help prey organisms escape from the predator. Now, in a very real sense, such adaptations represent a situation somewhat akin to that of a dog chasing its tail. This is particularly so when a given animal must be considered simultaneously as a predator as well as a prey for another beast. However, we may take the following as a simplifying assumption: Whatever adaptations serve the predatory prowess of an organism do not upset the devices that assure protection from its predators. If we accept this assumption, the problem can be disentangled, and our simplest approach is then represented by the question, How can behavioral adaptations protect prey organisms from excessive predation?

As a prey organism develops adaptations allowing it to escape its predators, one can expect the predators to evolve mechanisms that increase their efficiency for dealing with this prey. Spiders presumably evolved a sticky mucus for their webs, whose effectiveness was then vitiated by moths that developed hairs and detachable scales (Eisner et al., 1964).

This process of mutual adaptation and adjustment can continue indefinitely because the physical environment is inherently unstable. Thus a particular adaptation that at one point in time may favor a prey organism may at another time prove disadvantageous to that organism. Consequently the characteristics of a prey species need not be considered to be permanent, nor are those of the predator. The race between the adaptations of the predator for capturing the prey and those of the prey for escaping a predator may be viewed as a race whose finish line is constantly moved ahead of the contestants. In many instances it is the predator whose behavior we must focus on rather than the prey, since it is often the limitations of the predator's learning or sensory capabilities that must be seized on by the prey in evolving mechanisms for escape. Thus the development of mimetic coloration, though not in itself a behavioral adaptation on the part of the prey (or at least not necessarily so), nonetheless involves behavioral problems because mimicry can be successful only when it takes into account, in one way or another, the learning capacities and peculiarities of the predator. Mimicry that depends for its success on a particular pattern of coloration can hardly be expected to offer protection from predators that are physiologically color-blind. Similarly, the mimetic resemblances will also be of very little use when the abundance of either prey or predator is such as to reduce below the allowable minimum the number of contacts necessary for the appropriate type of learning to take place. We shall, therefore, find that our considerations often shift from the learning abilities or behavior of the prey to those of the predator.

A bit earlier we suggested that the two categories of explanations were not mutually exclusive. From a statistical standpoint, the fact that a smaller proportion of prey is killed at a low population density than at a high population density may be a function of the prey's being more widely dispersed through random or chance processes. Alternatively, there may be a "conscious" effort made on the part of a prey organism to maintain a certain distance from its conspecifics so as to reduce its density in any specific area and thereby reduce the likelihood of predators being attracted to that area. Brower (1958) has discussed this point in connection with the habit of certain butterfly larvae to locate themselves in a particular species of shrub. Presumably this is an adaptation that assures the avoidance of densities of individuals in a given bush of such magnitudes as to attract the attention of the predatory birds that prey on these larvae. We have in this particular instance of dispersal both elements of a statistical nature that are independent of the behavior of any individual organism and also explanations of a behavioral nature that depend on the individuals perceiving their distance from their conspecifics or on their ability to recognize appropriate perches or microhabitats. Territoriality, for instance, may reflect an adaptation of this sort and will be discussed in greater detail subsequently. (Note Vine, 1971, who balances out increased visual detection and pursuit by a predator with the benefits of flocking.)

Recognizing that there are no sharp lines to be drawn between these two levels of explanations, let us nonetheless focus on the behavioral level for the bulk of this chapter. Let us concern ourselves with protective mechanisms that enable a prey to escape from its predator or that at least reduce the likelihood of a prey being perceived by its predator, mechanisms whose effectiveness is a function of the behavioral capacities of either prey or predator. However, first a word must be said about the often-repeated notion that the principal device for protection from predators lies in nothing more than an adjustment of birthrates. The argument is that the greater the mortality in a given species, the greater is its birthrate, and that it is in the birthrate that the adjustment or compensation for mortality can be found. The hollowness of this argument, however widely it has been distributed in elementary and popular texts, should be apparent. It has been demonstrated that in at least some animals (and presumably in most) the number of young produced by given individuals is under as precise genetic control as any physical trait. That is, there are characteristic clutch or litter sizes not only for each species but for any individual (cf. Lerner, 1958). Such sizes being heritable, natural selection will favor organisms with a clutch size that allows them to leave a maximum number of offspring. Consequently, one might expect clutch sizes to increase independently of predation pressure. This increase does not, in fact, occur. It is known, for example, that among birds the number of eggs deposited in a clutch is well below the physiological limit that can be produced. (This problem has been carefully reviewed by Hutchinson, 1951, and Lack, 1954.) The flickers (*Colaptes* sp.), birds that normally may lay two, three, four, or at most five eggs in one clutch, can be induced to lay as many as five or six *dozen* eggs merely by the experimenter's successively removing eggs as they are laid. In the domestic fowl this situation is exploited by man for his own benefit.

The situation that obtains in fowl is very similar to that obtaining in almost all the gallinaceous birds, as well as in a great many others. Given the proper incentive, most birds can lay far more eggs than they actually do, and the physiological limit of egg production is normally not even approached under natural circumstances. We have, then, the fact that birds could lay many more eggs but do not, and the question we must answer now is why is this so? Part of the answer is suggested when one analyzes egg-laying habits of birds of different latitudes, as has been done by Lack (1954). Within the perching birds, for example, one finds a striking and significant increase in average clutch size as one goes from the equator northward. This striking increase in cluch size is not characteristic of marine ducks and certain others. A more intensive study of perching birds banded as nestlings brought out the fact that, within a given region, as the number of eggs in a clutch increased, the probability that the young would survive through an age of three months or more decreased. In fact, there was an optimum clutch size such that young from clutches of three, four, five eggs had a much greater probability of surviving than young from clutches of six, seven, eight, or nine eggs. And, similarly, young from clutches of one and two eggs, though having every chance of survival, nonetheless represented a smaller future population than those young coming from the slightly larger clutches. This factor was quite reasonably interpreted by Lack (1954) to mean that in passerines that feed their young the number of eggs laid is in large measure a function of the number that can be reared successfully without danger of underfeeding through having to divide a finite amount of food among too many hungry bills. The reason for the increase in clutch size with latitude is apparently the greater length of the spring days in the northern latitudes coupled with dense food supplies. The absence of this relationship in other birds is the exception proving the rule. Those groups in which the latitudinal correlation does not hold can be expected to be groups in which the young are not fed by the parents. Hence the survival of the young becomes less dependent on the activity or industry of the parent.

An equally convincing demonstration of the independence of birthrates and mortality is to be found in a study of the English gray heron (*Ardea* sp.), which normally maintains a population density that oscillates very little from year to year (Lack, 1954). Clutch sizes are relatively constant. In a particular year extreme conditions caused a crash in the population, leading to a striking decrease in the original population. Over the succeeding years numbers of this bird began to increase gradually until the original density that obtained before the crash was established once again. During this period of population increase there was no increase in clutch size. This fact demonstrates as clearly as anything can that clutch size, in this species, at least, is independent of mortality and is indeed adjusted to some variable—such as Lack (1954) has suggested—that is related to the optimum number of young that can be reared. The obvious explanation for the rise in population density in the case of the heron lies in the fact that there must have been a reduced mortality at low densities. Thus it should be clear that there are factors quite apart from natality levels that determine the number of young that are produced. With birthrate excluded, we can now turn to other adaptations protecting organisms from predation.*

PROTECTIVE COLORATION, CAMOUFLAGE, AND MIMICRY

An organism can escape from a potential predator in at least two different ways. First, it can exploit the limitations of the predator's motor or sensory capabilities and disperse or camouflage itself, in this way escaping the predator's notice. Second, prey organisms can

* MacArthur (1961) has argued that a species in choosing its geographic distribution will settle only where birthrate exceeds the inevitable (that is, density-dependent) mortality. If mortality at higher latitudes is greater because of more severe climates, longer migration routes, and so forth, then the clutch size must also be higher or the species would not occur there.

exploit the limits of the learning abilities of the predator by mimicking other organisms that the predator would normally be disinclined to take. For the present let us concern ourselves with the category of mechanisms that can be included under the rubric cryptic coloration and camouflage; dispersal and mimicry will be considered separately.

Cryptic Coloration and Camouflage

Cott (1956), in his monumental work, has reviewed the different forms that cryptic coloration may take and has discussed in great detail the optical principles on which crypticity is based. The important questions are the following: First, how can we determine whether an alleged instance of cryptic coloration really does represent a true attempt at camouflage rather than an instance of convergency or pleiotropy, an accidental and irrelevant by-product of some other physiological response? Second, does the prey organism show any awareness of the responses appropriate to cryptically marked animals? In other words, can the prey organism be thought of as showing any power of selection of the appropriate background or does the selection of matching background result from the selection by predators of mismatched prey?

In approaching the problem of the reality of an assumed crypticity, we can ask whether a predator against which the prey organism is presumably protecting itself is neurologically capable of perceiving the stimulus patterns exploited by the prey organism. Elaborate forms of color camouflage, for example, could be of very little value if the chief predator against which protection was sought was an animal that was absolutely devoid of color vision. Unfortunately, questions concerning the neurological capabilities of animals are answered only with great difficulty. Walls (1942) has reviewed some of the types of tests necessary to establish the presence or absence of color vision in a great number of vertebrate and invertebrate forms. He comes to the conclusion that, in point of fact, very few workers have exhaustively eliminated all possible alternatives when considering whether color vision does or does not exist. In all too many instances, a presumed discrimination on the basis of color can be explained in terms of differences in saturation or brightness or intensity or some other fully irrelevant cue.

Even where a neurological capability exists in terms of the appropriate peripheral mechanisms there is no assurance that the animal involved is functionally capable of the required perception. It is, for example, not too difficult to establish that particular wavelengths of light will cause a small group of receptors in a retina to fire. The techniques of Granit (1955), who has recorded impulses from single elements of the retina, have often been adapted for studies of this sort. With such techniques it is possible to show that stimulation of the retina of a cat with various pure spectral colors does result in the type of dominator-modulator curve that, according to Granit's work, is characteristic of organisms with bona fide color vision. However, it is extremely difficult to train cats to make a discrimination on the basis of wavelength alone when all other cues for the discrimination are eliminated. This factor suggests that the cat, from a functional standpoint, may be quite colorblind whatever its peripheral capabilities may be.

Here a common difficulty is intruded, raising the problem of response and stimulus equivalence. The difficulty stems from the fact that given a particular stimulus an animal may show a number of responses. How can one determine which of these responses really represents the critical measure of the animal's reaction to the stimulus? Further, let us say that the stimulus provokes a change in the amplitude, as well as in the frequency, of a nerve discharge. Are these two variables to be considered as equivalent? Or, passing to another level, let us assume that there is a certain change in the discharge pattern that results from the stimulus and that there is a gross behavioral change as well. What is the relationship between the gross behavioral change and the change in the neural pattern of discharge? Are these two kinds of responses commensurate in the sense that one inevitably means the other or are they perhaps independent reactions lacking a causal link between them? The experiment with the cat is an illustration of this difficulty. We do not know whether the cat was indeed capable of seeing color, but "unwilling" to learn the required discriminations, or was essentially blind.

Ultimately the problem of commensurability can be raised on yet another level. Suppose that we have an animal that shows itself neurally capable of responding to certain classes of stimuli. Suppose further that this organism is functionally capable of responding. We then have to ask whether or not the response elicited in the lab is of any biological significance. There are a number of instances of behavior, with which we have become familiar in recent years, that offer no demonstrable selective advantages to the animal and that also have never been observed in animals under conditions less artificial than those of the laboratory. Koehler (1956), in his experiments on the unnamed number concept in birds, has rather convincing data to show that several species of birds are able to count, or act upon certain quantities, up to five or six quite regularly, proceeding, in a few rare cases, to numbers as high as seven. Thus a bird can be trained to select from a pile of grain five kernels when shown a card with five symbols on it. Alternatively, it may be taught to proceed along a line of dishes with variable amounts of grain in each one, removing lids and eating the grains, until a number of grains has been consumed that corresponds with the original number of patterns displayed on the card. It is clear that the animal really has developed some type of unnamed concept and is not simply utilizing a rhythm-counting technique. What conceivable use such an ability can have to a bird in the wild remains to be guessed. It is quite possible that the ability to form unnamed number concepts is merely an expression of a more general intellectual ability that does indeed stand the bird in good stead. The primary fact is that we have again demonstrated a particular form of behavior under one set of conditions that allows no conclusions to be drawn, or at most allows only very limited conclusions, about the animal's performance under other sets of conditions. In general, then, this problem of commensurability that exists on the three levels mentioned represents one of the more critical problems that must be faced in discussions attempting to relate the behavior of an animal to its general ecology.

Let us turn now to those situations under which there is no question but that the form of an animal does serve a cryptic function. It is then of interest to know whether the prey organisms are able to exercise some kind of choice with respect to the appropriate background or, alternatively, whether there is a process of selection that assures the distribution of the different patterns on matching backgrounds.

The work of Sheppard (1959) and Cain and Sheppard (1954) and Goodhart (1958) on the snails *Cepea nemoralis* and *Cepea hortensis* in England, for instance, has shown that the proportion of snails of different colors that can be found varies with the season of the year. In the spring, when the ground is green, the proportion of snails with yellowish and greenish tints in their shells is considerably higher than in winter, when the colors tend to be predominantly of the reddish sort that match well the bare hard soil of Cambridgeshire.

Conversely, the proportion of predated snails or empty shells of different colors that are found at the thrush anvils (the stones on which the avian predators break their prey) varies with the season, too. There the shells of the improperly matched color predominate at a given time of the year. It would not be difficult to imagine a situation in which the proportions of different color types were fairly constant. The fact that investigators would then find green or yellow snails predominating in the spring and summer and pink ones in the winter would be attributed to the fact that the contrasting color was more readily seen by the thrushes (*Turdus* sp.) and thus selectively removed. The other possibilities are that the snails are somewhat capable of matching their own color with that of the background or that a polymorphism exists that assures that a different proportion of the different color types appears at different seasons of the year. Differences in the activities of the snails of the two different color types have also been suggested, raising still further complications, though we need not consider them at the present (note Wolda, 1963).

To gain further insight into the nature of this instance of crypticity, data about the sensory capacities and behavior of these snails are required. Another example may be cited that illustrates even more cogently the need for such information. Bergman (1955) has shown that the downy young of certain *Hydroprogne* and *Sterna* species (terns) possess a color matching that of the substrate on which they are hatched. Beaches of different colors thus have colonies whose downy young differ in color. Whether these birds show a pronounced *Ortstreue,* or locality imprinting, or whether other perceptual processes are involved in the recognition by the adults of the appropriate breeding site is unknown. Until this fact does become known, the possibility of a random production of color type and a nonrandom selection each season by predators cannot be discounted.

A beginning in studies of background recognition has been made. For instance, Sargent (1968) has data that indicate that certain moths do select the correct background on which to alight. Painting the circumocular scales does not abolish their selectivity, so presumably something other than mere brightness matching is involved. Unfortunately, the previous history of the moths used in these studies was not known, so other interpretations remain possible, too. And in stick insects it is now known that the visual mechanisms are particularly sensitive to just those elements that characterize their preferred domain (Jander and Volk-Heinrichs, 1970).

The Perceptual World of Other Organisms

The problems already discussed point to the necessity of knowing the perceptual world of other organisms. How is this knowledge possible? Even in the case of our own species this problem has given much difficulty to psychologists and philosophers alike. Many rational arguments have led to the solipsistic viewpoint that admits that it is impossible to know of the perceptual world of any organism other than oneself. Some investigators, such as von Uexküell (1921), would have us infer the perceptual world, or *Umwelt,* of an organism from the structure of its sensory apparatus. While this view has certainly much to recommend it and has had the advantage of focusing on the fact that different organisms are indubitably sensitive to different categories of stimuli, it nonetheless leads to a fair degree of anthropomorphism and, to this extent, is apt to lead one into error. It is only an operational viewpoint, which defines an animal's perceptual world in terms of the animal's responses, that can be seriously considered by an empiricist. But the operational viewpoint alone does not solve all difficulties. How does one measure responses? How does one determine the commensurability of the different measures? These are problems with which the operationalist all too seldom comes to grips.

The problems alluded to above are aggravated by differences between vertebrates and invertebrates insofar as the integrative processes of their nervous systems are concerned. Since the nervous systems of vertebrates and invertebrates do differ in a number of fundamental ways, it is clearly reasonable to suppose that the techniques that have been evolved for determining the sensory and perceptual abilities of one animal cannot be applied without modification to another, particularly as one shifts from birds to snails.

Some of these differences in the mechanisms of nervous integration have been summarized by Vowles (1961). For example, he points out that in insects most motor neurons innervate all the motor effectors. This process is in striking contrast with the situation in the vertebrate nervous system where, by and large, a small number of motor neurons stimulate a single effector and no other. This is particularly true in the case of the effectors that are used for highly refined movements. The integration of gross muscle responses occurs not peripherally but centrally. Gradations in the responses of a whole muscle made up of many motor units are effected by variations in the number of component units that are stimulated at a given time. In the invertebrates, the many fibers innervating the effector may produce localized effects rather than a more general depolarization that extends along the entire motor unit. A single motor fiber thus may be made to contract only along a very small portion of its entire length. In addition, the various motor neurons that project onto the motor unit may produce rather different effects, some causing mild and some intense contractions, some inhibiting contraction, and so forth. Integration, in other words, occurs at a peripheral level. This method of integration has consequences for the perceptual abilities of these organisms. It can easily be seen in the fact that invertebrates are, by and large, much more limited to serial perceptions or perceptions of successive waves of stimulation than are vertebrates, which have integration processes taking place centrally, utilizing simultaneously blocks of neurons. The processes of integration among invertebrates have been clearly illustrated by the analyses of the response of mantids (*Mantis* sp.) to prey organisms by Mittelstaedt (1962), as well as by the elegant studies of Young (1961), Boycott (1954), and Wells (1959) on the perceptual systems and learning abilities of the octopus (*Octopus* sp.). (See also Gilbert and Sutherland, 1969.) Visual discrimination in the octopus, in accordance with the views of these workers, is based upon a grid (as depicted in Fig. 1–1) along whose two axes the number of units that are stimulated is a function of the linear extent of the object in that particular plane. Similarly, tactual discriminations are based on the number of adjacent papillae in the arms that are stimulated. Different patterns may be etched into the surface of Perspex cylinders, and yet these cylinders may be judged alike or different depending purely on the similarity or difference of the grove–surface ratio.

The lesson to be drawn from such analyses as these is that crypticity could indeed exist with respect to the prey of animals such as the octopus and pass unnoticed by man because the perceptual and sensory processes characteristic of man differ so radically from those of the octopus. Thus a sensory analysis, of perceptual mechanisms along the lines of those that have been carried out for the octopus or the mantid, would appear to be a prerequisite for any determination as to whether or not a given form or pattern is truly cryptic.

A further complication is introduced when the patently different nervous system of an invertebrate must be attuned to the behavior of a vertebrate. Roeder (1959) has suggested that the

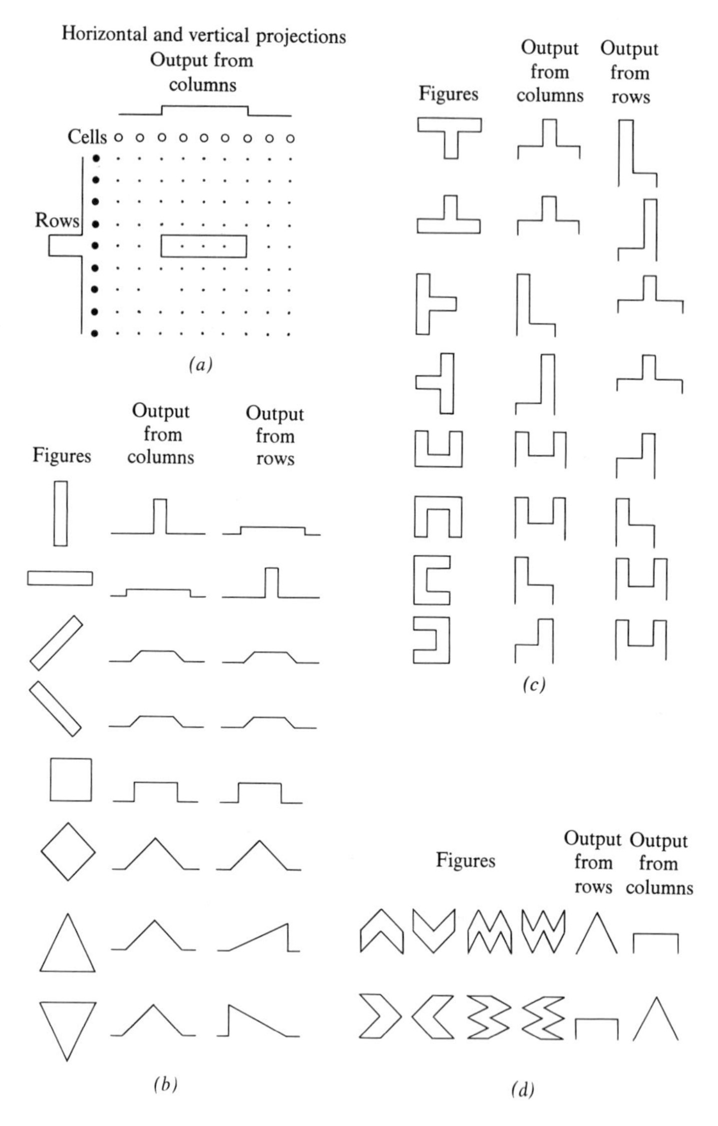

Fig. 1–1 Sutherland's hypothesis of shape discrimination in the octopus. (*a*) The small dots represent the retinal array with a horizontal rectangle projected onto it. Open circles at the top represent cells specific to each column, and solid circles at the side represent cells specific to each row. The cells for the rows and columns are supposed to be connected to distinct outputs (Sutherland, 1957). (*b*) The outputs produced by various rectangles, squares, diamonds, and triangles (Sutherland, 1958). (*c*) Mirror images (Sutherland, 1960a). (*d*) Figures that should be confused (Sutherland, 1960b). Figure and references after Young, 1961.)

small size of insects relative to their vertebrate predators and the slow conduction of their unmyelinated fibers require that a disproportionate amount of their neural material be allocated to large and more rapidly conducting fibers. Only in this way can response latencies of equal orders of magnitude to those of vertebrates be achieved. The result is yet to constrain further the sensory and perceptual parts of the invertebrate nervous system.

In the case of vertebrates, the situation may be less complex in that most mechanisms for neural integration appear to be rather similar to those that operate in man. But still there are three main kinds of differences that separate man from other vertebrates and introduce difficulties. First, sensory modalities that some animals possess may be wholly unrepresented in man, causing him to totally overlook the existence of cues obvious to other beats. The Gymnotidae (a family of fish), for instance, show extreme sensitivity to minor fluctuations in the electrical fields surrounding them. Lissman (1958) has shown that it is possible to train these fish to differentiate solid from hollow cylinders purely on the basis of the differences in the distortions created by these objects in the electrical field produced by the fish. This type of "radar" detection undoubtedly serves both to guide the animal in its movements and to aid it in finding appropriate food objects, yet this is a modality for which we have nothing comparable. We can only now begin to guess the characteristics that the prey of the Gymnotidae would have to possess to be cryptic. Apparently the electric discharges also play a role in social and territorial behavior (Black-Cleworth, 1970).

Second, it is possible for an organism to differ from man in the extremes of sensitivity it possesses within particular modalities. Bees, for instance, are sensitive to ultraviolet portions of the visual spectrum to which man is largely blind, and man, in turn, is responsive to red portions of the spectrum that bees cannot see.

Finally, there are unquestionable differences in the degree to which patterns of stimuli are perceived as "wholes," or *Gestalten,* or recognized by virtue of specific sign-stimuli. These differences in the manner of perception are reflected by the studies of human perception by the Gestalt psychologists, on the one hand, and the studies of releasers and sign-stimuli by the ethologists (cf. N. Tinbergen, 1951; Lorenz, 1937), on the other.

In many birds, it is characteristic that they recognize biologically appropriate objects by reference to a very limited portion of the total stimulus character of these objects, which are known as releasers or sign-stimuli. Lack (1939) has shown that a male robin (*Erithacus rubecula*) defending its territory will display as vigorously to a small red feather mounted on the end of a wire as it will to a very realistically stuffed male robin. All that is really relevant to the territorial male is the patch of red at a particular height. Similarly, as Noble (1936) has shown in the flicker, *Colaptes auratus,* the recognition of the opposite sex may based on the presence or absence of a dark moustachial streak and nothing else. Again, perceptual processes in recognition are limited to a very small aspect of the total stimulus situation. The male simply failed to recognize his mate so long as she bore the small amount of black by her head. When the streak was removed, she was promptly reaccepted. The type of sensory integration involved in the perception of such sign-stimuli is certainly of a totally different nature than that which is involved in Gestalt perception, a process that focuses not on specific isolated features of the organism but on the whole configuration. Considerations as to which type of perceptual process is involved in a particular instance are very much relevant to the problem of deciding whether or not an organism's pattern represents an attempt at crypticity. What is cryptic for a predator whose perceptions are organized on Gestalt outlines may not be cryptic for an organism that is responding to specific and restricted sign-stimuli.

The difference between the two classes of perception, Gestalt and sign-stimulus, that we have alluded to, finds its corollary in the functional characteristics of the central nervous system. It is well known from the works of Lashley (1949) and Beach (1951) that a large range of responses made by mammals to sexual stimuli, or in form discrimination, is scarcely interfered with by the removal of major portions of the cortex or by the interference with any number of sensory modalities. Specific sensory deficiencies do not inhibit responses to complex stimuli. It is probable that animals that rely more on sign-stimuli, such as fish, birds, or insects, will show quite the reverse tendency. In general, then, one would expect to see a difference in the structure of the various nervous systems paralleling the differences in the kinds of perception that have been noted.

Even with a full store of information on sensory capacities, perceptual differences between animals complicate our studies. Most studies of sensory capacity depend on tests that require the learning of differences. Work done in a variety of labs has shown that conditions defined as "reenforcing," i.e., increasing the probability of a response being repeated or a discrimination learned, vary in effectiveness depending on the context in which they are presented. Many times this is obvious: An icy lemonade is not likely to appeal to a frostbitten child on a bitter winter morning. It is less obvious, but still reasonable, that rats (*Rattus* sp.) learn a particular pathway through a maze more readily with a drink as a reward than with food. A variable path, on the other hand, is more easily established with a food reward (Oatley, 1970). In the usual world of rats, the location of water is fixed, while food is found in different places. This apparently influences the effectiveness of food and water as rewards in the different learning contexts. This principle applies to humans, too, apparently. The differences in IQ scores made by children of different social backgrounds may reflect no more than differences in the incentive value of the traditional rewards (Zigler, 1967). The reverse situation also obtains where the effectiveness of a cue varies with the nature of the reward (Garcia et al., 1968; Garcia, 1971).

Differences in discriminative ability also appear depending on whether the training involves successive or simultaneous presentations of the discriminanda (Bloomfield, 1968). Finally, "context," which has not been rigorously defined, apparently must include endocrine states, too: Migratory birds viewing a particular star pattern will head south or north depending on whether the photoperiod to which they had been subjected had been simulating spring or autumnal conditions (Emlen, 1969).

The foregoing examples make some kind of intuitive sense. The normal life history of the animal in question is somehow consonant with how it treats those stimulus contingencies that we call cues, rewards, and the like. Thus the statement that context, the totality of the situation, defines the important contingencies does not seem any bar to coming to an understanding of perceptual abilities through thoughtful experimentation. Unfortunately such optimism must be tempered by the consideration of other examples that do not yet seem to make sense. Young ducklings, *Anas platyrhynchos,* discriminate patterns in a maze-learning context that they fail to discriminate in an imprinting context, or vice versa (Klopfer, 1968); in some mammals the character of the blindness that results from certain cerebral insults depends on the stimulus used or the response demanded (Schneider, 1969). Reasonable explanations for these and similar observations can be expected, though the ingenuity of the experimenter may certainly be taxed and his patience frustrated. But so long as the explanations are still wanting, no amount of hopeful anthropomorphizing will truly allow us to see the world as other organisms perceive it.

Mimicry and Its Contingencies

The foregoing sections dealt with escape mechanisms whose effectiveness depended largely on sensory or perceptual limitations of the would-be predator. Here, let us consider adaptations whose effectiveness is even more closely attuned to the complexities of the central nervous system of the predator, specifically, on limitations in their memory as well as perceptual and learning abilities. These are the adaptations termed mimicry, whereby an organism eludes its pursuer through a misleading resemblance to another.

The evolutionary processes that produced mimicry in the first instance are not within the scope of this book. The treatment of this theme by Fisher (1930) is not likely to be improved upon here. Some of the experimental evidence as to the reality of mimicry and the selective advantage it affords is summarized in Sheppard (1959) (and see Brower et al., 1967, and Wickler, 1968).

There are several behavioral problems that intrude themselves into considerations of mimicry. For example, if a predator discovers that which appears to be a single type of prey sometimes is palatable and sometimes is not, will he avoid this prey or not? Next, to what degree must model and mimic differ in palatability or in abundance for the mimic to benefit from its resemblance to the less palatable or unpalatable model? Can a high degree of palatability on the mimic's part be compensated for by a relatively much greater abundance of the model? To what extent are intensity and frequency of negative reenforcement related? Further, what results are to be expected if an encounter with a model proves fatal to the predator? Could mimicry then arise? How frequently must a correct discrimination by the predator be reenforced within a given period of time in order for the correct response to persist? And, finally, how does social or communal feeding behavior influence individual learning rates and responses? These are clearly interrelated problems, but we shall, for expository purposes, attempt to view them independently. In each case we shall cite a biological example, discuss relevant psychological experiments, and, finally, suggest further studies.

Relative palatability

De Ruiter (1952), using caterpillars of the genera *Ennomos* and *Biston* as prey and European jays (*Garrulus* sp.) as predators, found that the latter often failed to recognize these stick insects as food. But, once they had been induced to peck at one larva, the jays eagerly devoured as many more as they could find. What is more, they then also began to peck at twigs from the caterpillar's host trees, to which the larvae bear an uncanny resemblance. This was an act in which the jays had never formerly indulged. De Ruiter points out that in nature these stick caterpillars do not congregate but are widely dispersed. Hence the probability of a predator finding an edible larva rather than an inedible twig must be fairly low, and the presumption is strong that this fact plays a major role in the protective value of the caterpillar's habit of dispersing. Should a "mistake" by the jay be more seriously disadvantageous than that resulting from mistakenly pecking a twig, as when feeding opportunities are so limited as to make wasted movements mean a loss in food or when the mistake may prove fatal, one might expect that the mimic could become relatively more abundant. The deadly coral snakes (*Micrurus*) and their presumed mimics represent an extreme instance of the situation in which a large difference exists between the results of a correct and incorrect response by the predator. The specific question to which this example is addressed is, How can mimicry arise if the unavoidable errors in learning prove fatal to the learner? The assumption, of course, is that a predator will learn to avoid coral snake mimics only after having had an unfortunate encounter with a coral snake itself. There are good reasons to believe that avian predators, at least, will either do rather well by attacking a coral snake (the snakes appear to be acceptable food) or else they will not live to repeat their attack. Hecht and Marien (1956) have recently shown that this particular problem was somewhat misapprehended inasmuch as many of the alleged coral snake mimics are themselves slightly poisonous. The general problem remains, though better examples might be found. In any event, the data of Klopfer (1957) suggest that in such instances a form of observational learning may operate such that a congener of a stricken predator will profit from his unfortunate fellow's encounter. In these experiments, ducks (*Anas platyrhynchos*) were trained to avoid an electrically wired dish through which a painful shock could be administered. Other ducks that had never been shocked themselves but had observed their congeners being shocked also subsequently avoided the dish. Social augmentation of learning has been noted in other animals and situations, too (Gans, 1964; Turner, 1964).

Psychologists, of course, have long been concerned with the related problem of incentive magnitude and learning rate. One review of the subject (Pubols, 1960) tends to minimize the effects of differences in quantitative variations in incentive magnitude; however, an effect on extinction rates was shown. Significantly, these results are dependent on the particular measure of learning that is used. With time-dependent measures there is indeed a demonstrable effect of incentive magnitude.

Also, the greater the cost of an error, at least in some situations and for some animals, the more conservative is the discrimination (Duncan and Sheppard, 1965).

The fundamental issue is whether there exists some threshold difference in palatability of prey that suffices to promote mimicry or whether the effect of palatability differences is continuous; any difference will do, but the greater the difference, the greater is the effect. It may be that different answers will be found for different predators.

Relative abundance

An analysis to show the role of the relative abundance of mimics was attempted by Brower and Brower (1960), using captive starlings (*Sturnus vulgarus*) as the predators and painted mealworms (*Tenebrio* larvae) as the prey. Some of these mealworms were rendered unpalatable by treatment with a quinine salt. Through a comparison of the experimentally determined learning curves of birds confronted with different proportions of edible and inedible mealworms, their results show, in general, an increased learning speed with an increase in the proportion of the inedible "models." Yet, it is clear from other work that the interval between tests may be a crucial variable in determining learning rates, and even in determining whether learning will occur at all.

Spaced trials are generally more effective in establishing a simple stimulus–response sequence than massed trials. The mimic's success depends on the predator's learning that the model is distasteful. Thus, where the models and mimics are so dispersed as to allow for the lapse of only the optimum period of time between trials, a greater selective advantage accrues to the mimic than in situations where model and mimic are so concentrated as to provide opportunity for massed trials. An interesting example has been provided by the work of Prop (1960), who has shown that naive birds are initially frightened by the jerking displays of sawfly larvae (*Diprion*). Later, some habituation appears to occur and aggregations of sawfly larvae will then elicit feeding responses. As the unpalatability of the larvae is related to the degree to which they aggregate, the would-be predators learn to avoid this distasteful prey. Were palatable mimics of these sawfly artificially produced and added to the aggregations, we would predict that they would be at a severe disadvantage as compared with mimics whose models were more widely dispersed.

Frequency of contacts

The next of the general problems alluded to earlier concerns the importance of intermittent reenforcement. What is referred to is the well-known fact that a rat (*Rattus* sp.) in an automated device for training its occupant to push a lever or perform some similar task (called a Skinner box) will show more stable learning if the rewards do not follow every correct response but are provided intermittently. The reward schedule may be fixed or not, and, indeed, in some instances the randomized schedule would appear to promote the most stable learning.

Thus there must exist a quantitative relationship between the relative and absolute abundances of the model and mimic as well as the differences in their degree of palatability. The less the difference between the two in the latter respect, the greater must be the difference in their respective densities. A difference in a palatability so extreme as to produce a fatal result in the event of an error by the predator (fatal to the predator, or course) is mitigated by the existence of an ability, possessed by some predators, for observational learning (see p. 22). Finally, since spaced trials most readily promote the learning essential for making mimicry advantageous to a potential mimic, natural selection will favor low absolute densities of both model and mimic (relative to the predator), as well as dispersion. Thus, spaced rather than massed trials should be the rule in natural populations.

One advantageous circumstance that biologists can exploit is the following: Some animals may sequester toxic or distasteful substances from their food. Danaine butterflies (such as the monarch, *Danaus plexippus*) assimilate cardiac poisons from the milkweeds upon which they feed (Brower, 1970). This, in turn, may have unpleasant effects on their predators. The poison content of the milkweeds varies both qualitatively and quantitatively; hence not all monarch butterflies are equally distasteful. This has given rise to what Brower (the discoverer of the phenomenon) terms automimicry. The presumption is that all monarchs will benefit from a few of their kind being distasteful. Since both degree of palatability and proportions of unpalatable individuals may vary independently, it becomes rather interesting to establish if there is an optimum combination for the maintenence of the protective advantage.

Observational learning

The fourth of the problems concerns the role of social interactions. Klopfer (1957) has, as already noted, described an instance in which there existed an ability to learn an avoidance response by the observation of the responses of a congener. From more recent work, however, it has become apparent that what was then called observational or empathic learning actually may involve several rather distinct stimulus situations, each having rather different effects. First, the observing animal may have its attention directed toward a particular prey or other source of food by the activity of another animal. This phenomenon constitutes what Thorpe (1956) has called local enhancement. Hinde and Fisher (1952) and Fisher and Hinde (1950) have suggested this process as the explanation for the spread of the cream-stealing habits of the British tits (*Parus* sp.). Originally confined to localized areas in the British Isles, the bird-caused disappearance of cream from milk bottles left on the doorstep has now become common throughout much of Britain and continental Europe.

Local enhancement, however, is best considered as a special case of social facilitation (Thorpe, 1956) or "contagious behavior." The classic example of this process is the satiated hen that recommences eating when placed together with a hungry, feeding flock. What seems to be involved is an association between the perception of certain motor acts performed by others and particular affective states, which, in turn, elicit the same motor acts in the observer. These effects are transient in both the foregoing cases and require the continued presence of the observed companions. Where the observer copies an act of a performer and does so under conditions leading one to suspect a degree of "self-consciousness" on the observer's part, Thorpe (1956) speaks of true imitation. Operationally, this category would seem to pose a sticky problem indeed, though there are good reasons for presuming the existence of insightful imitation on the part of many animals. This problem has been reviewed by Thorpe (1956) and will not be further explored here.

A case has been made, however, for there being objective criteria whereby visual imitation in Thorpe's sense can sometimes be recognized. These criteria were described in a paper by Porter (1910). Porter allowed a bird to learn to open a box that could be opened in any one of several different ways. When the first bird had mastered a particular technique for opening the box and used it to the exclusion of other techniques, he was replaced by a second bird but allowed to observe his replacement's efforts at solving the problem. Whenever the second bird developed a different technique and the first bird subsequently modified his own methods to more closely conform to those of bird 2, imitation was said to have occurred. In one sense, this behavior becomes a criterion of insightful imitation only because no other explanation is evident for the change in bird 1's behavior. To the degree that this is the case, the concept of imitation must still be considered to have been defined negatively. A similar situation was described by Meischner (1964), who also found that birds can associate specific visual patterns with particular motor responses. The examples of acts that fall into this category are so numerous (for example, Harlow, 1959) that they justify much more study of this problem. Some further points will be considered in Chapter Two.

Finally, much learning by observation can be attributed to a form of conditioning. Assume that an animal emits an alarm signal while in contact with a particular stimulus. If this signal can serve as an unconditional stimulus for a naive observer, that is, if the inexperienced observer will react in a predictable fashion to the alarm signal, the observer's response may become associated with the stimulus that elicited the alarm signal from the observed animal. Similarly, if a given stimulus elicits a feeding response in one beast, a second animal may form an association between the feeding behavior and the relevant stimulus without himself ever having fed. This has been called secondary conditioning (Klopfer, 1959a, 1961a and b) to distinguish it from the conditioning type 1 of Konorski (1948). It may be distinguished from social facilitation by its relative stability; once established, the response of the observer persists in the absence of the performer. A striking example is provided by Rothschild and Ford (1968), where a warning call from one starling (*Sturnus vulgaris*) led another to reject food.

These considerations are relevant both to the special problem of mimicry and to the more general problem of the establishment of food preferences. The latter will be considered in greater detail in Chapter Two, but it will be well to discuss here certain experiments illustrating the difficulties confronting an organism that must learn to distinguish between palatable and unpalatable food or safe and dangerous prey.

If a single greenfinch (*Chloris chloris*) is trained to take a particular kind of food and to avoid another, he may learn the discrimination in as little as one trial (Klopfer, 1959a, 1961a and b). Learning is very rapid. The presence of untrained companions tends to slow the learning process somewhat, probably by virtue of the distracting effect of the companions, but the discrimination can still be readily acquired. If, however, a trained greenfinch observes an untrained bird making some errors during the latter's preliminary trials, it may begin making errors itself (Fig. 1–2). Described anthropomorphically, it is as if the trained bird, seeing a naive companion feeding upon the forbidden fruit, thinks, "Aha, it must be

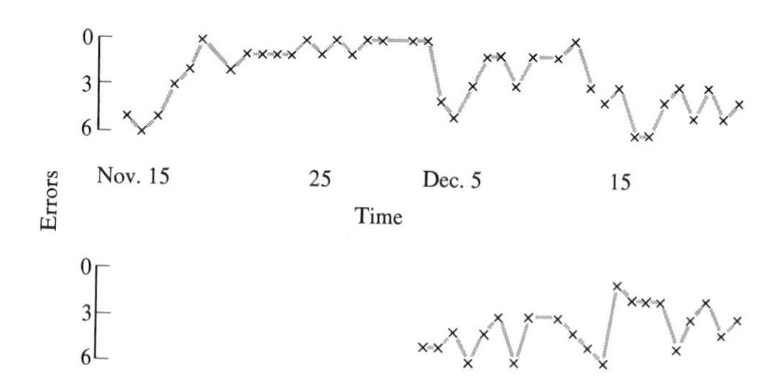

Fig. 1–2 The upper curve depicts the performance of an adult male greenfinch who learned, after four days, to discriminate between two patterns, making no more than two errors in any one day (each X represents a series of trials held on successive days). After about two weeks, an "observer" male is trained (lower curve). At this time the original "actor" appears to "unlearn" the discrimination. Thereupon, the two birds oscillate erratically, apparently interfering with one another's learning (Klopfer, 1959a).

good, after all." Two birds may then repeatedly learn and unlearn a discrimination, cycling out of phase with one another. Careful analysis reveals that what is probably happening is that the feeding behavior of one greenfinch is an unconditional stimulus potent enough to overcome the effects of the second bird's own previous experience with the unpalatable food. Such an interference with the normal course of learning would be highly maladaptive in many cases, so it is of interest to note that in at least one species with highly explorative habits, the great tit (*Parus major*), this kind of interference effect cannot be observed (Fig. 1–3). The greenfinch presumably has sufficiently conservative food habits as to be unlikely to attempt to use new foods that might prove undesirable. It would appear, then, that mimicry would not be expected to arise in forms whose principal predators showed a "greenfinch" learning pattern, while it would be most common in forms preyed upon by birds of the "great tit" type. A test of this hypothesis could readily be made with rhesus monkeys

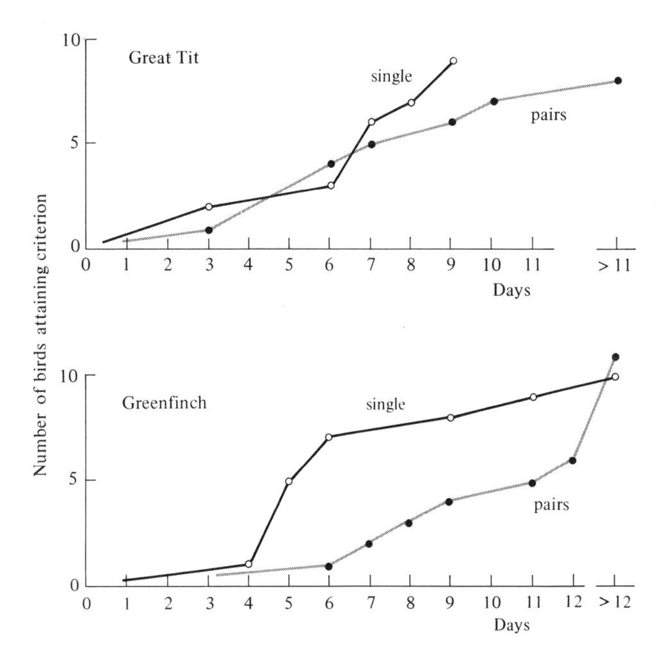

Fig. 1–3 The difference between greenfinches and great tits in the effects of a partner on learning rates. The single line represents the number of days that various singly trained birds required to reach a criterion level of performance. The double line represents this for pairs of birds trained together. Among the greenfinch, single birds generally do better than pairs; this is not true for tits (Klopfer, 1961).

(*Macaca rhesus*), for Riopelle (1960) has shown that observers learn more readily when the demonstrator errs (makes an unrewarded response) than when he makes a correct, rewarded response.

Additional studies of these sorts have raised some ancillary questions. Tolman and Wilson (1965) find that a certain degree of social interaction is prerequisite to facilitory learning effects; the observations of Miyadi (1959) and Itani (1958) suggest that the directionality of learning (in macaques, *Macaca*) may be influenced by social status—peers learning from peers or superiors but not from subordinates. Feldman and Klopfer (in press) suggest that this is also true for lemurs (*Lemur*). The best evidence on this matter comes from Chesler (1969), who showed that kittens learned more readily from their own mother than from another cat (*Felis* sp.). Whether this is due merely to differences in attentiveness or to other factors remains an open question.

Search images and oddity effects

The concept of "searching image" was introduced by Lukas Tinbergen to explain certain discrepancies between the abundance of a prey organism and the numbers actually taken by their avian predators. At low densities, a new prey species may at first be altogether overlooked only to become especially esteemed later. This was assumed to result from the fact that once a new constellation of characteristics is discovered to mean "food," it is selectively sought (Croze, 1970).

A complicating factor has been noted by Mueller (1968), however, in that some predators seem attracted by the odd or most divergent member of an assemblage of prey. Thus the more cryptically marked animal could be at a disadvantage if it is least like its fellows. Mueller is studying the complex interaction of crypticity, search image, and oddity preferences. He points to the significance of oddity selection to population regulation, with this assuring removal of mutants, sick and injured animals, or those not in their normal habitat.

Degree of resemblance

A study by Sexton (1960) has shown that the degree of resemblance required between model and mimic for mimicry to be effective may depend on whether both model and mimic are simultaneously present. In his experiments, the presence of both model and mimic at the same time required the mimic to show a higher degree of similarity to the model than when the mimic was present alone. Thus the results of mimicry experiments can be expected to vary according to whether a simultaneous or successive presentation technique is used, a point not usually noted by experimenters. Sexton et al. (1966) found that in the anolis lizard (*Anolis cardinensis*) the amount previously eaten (not surprisingly) influenced the subsequent response. A review of the vast literature on discrimination behavior would undoubtedly provide other clues concerning when mimicry might or might not be expected to arise.

Delayed reaction

In some instances a noxious organism may not induce an immediate response. The toxins of certain fungi produce their effects only after some hours, or even days. Wherever such situations arise, the possibility for mimicry to arise can exist only where the prospective predator's powers of memory are sufficiently good to allow it to retain the image of the noxious model for a period of time equal to the latency of the noxious effect. Macura (1959) has shown that reactions delayed up to 30 minutes are possible in tawny owls (*Strix aluco*). His birds were simply shown under which of several dishes food was placed, and the time over which they could recall the position of the food was recorded. This delay, however, represents a duration of time longer than that which most subprimates can sustain. Relatively little is known about the delay possible in forms other than mammals, the tawny owls excepted. The search images of the tits (*Parus*) of L. Tinbergen (1960) changed slowly with changes in the prey density, though often with a lag of several days or weeks. This lag suggests a rather stable memory. Bees, certainly, are capable of phenomenally long delayed reactions, but here, too, precise parameters have yet to be established.

The measure of memory is influenced by a variety of contextual and internal factors. The folk belief that an intoxicated person can more readily recall certain actions performed during a previous drunken binge than when he is sober appears to have a basis in fact (Goodwin et al., 1969). Nor need memory be conceived as a unitary phenomenon. Electroconvulsive shock administered directly upon a single-trial learning episode is presumed to induce a retrograde amnesia, but this is apparently true only if response latency is the criterion. If changes in EKG (electrocardiogram) are used as measures, there is no evidence for forgetting (Hine and Paolino, 1970). Finally, the allowable delay may vary with the modality. Garcia (1971) has shown how (in rats, *Rattus* sp.) tastes but not smells or sights can be associated with a food that long afterward produces malaise. A visual cue can certainly be associated with a painful foot shock and will be appropriately used thereafter. Neither the appearance of tainted food, however, nor even its odor, can serve as a reminder of a delayed emetic effect. Only taste serves. The generality of this phenomenon among animals is unknown (cf. Kalat and Rozin, 1970).

Confusing the predator

The myth of Proteus is highly credible in at least one respect. Rapid changes in form make objects we wish to capture disconcertingly difficult to seize. A priori, there seems little reason for doubting that most vertebrates would be similarly disturbed. If true, a significant selective advantage might be expected to accrue to organisms capable of what have been termed "Protean" displays (Chance and Russell, 1959).

In our previous examples, protection was gained either when a palatable form mimicked an unpalatable or dangerous form (Batesian mimicry) or when it camouflaged itself (cryptic coloration). When two unpalatable forms resemble one another (Muellerian mimicry), protection may be gained because then a predator need learn fewer cues. An instance of this factor is afforded in a study of Blest (1957) in which it was demonstrated that the sudden appearance of bright concentric circles (similar to the marginal eyespots of many Lepidopteran (butterfly) wings) was frightening to birds and inhibited attack. Coppinger (1970), too, found that novel stimuli often failed to elicit attack.

In all these cases, however, there is a regularity or predictability about what is found. The relative proportions of models and mimics, as has been suggested earlier, are fixed within specific limits determined by the learning abilities and characteristics of their predators. Polymorphism, particularly of the type where only a proportion of a species (e.g., one sex) possesses the features of a mimic, may increase the complexity of the predator's problems to the advantage of its prey.

This effect is equally true of polyethism, displays or behavior patterns that are distributed nonuniformly through a population. The ability of some ducklings to learn rapidly responses to either auditory or visual stimuli, but not always to both, is an example of one type of polyethism (Klopfer and Gottlieb, 1962a and b). Presumably, the ability of some birds to be most efficient in one modality and others in another is under genetic control in a manner analogous to the morphs of certain butterflies, though this presumption has yet to be experimentally confirmed. Another kind of polyethism can introduce an even greater complexity. This is the case when the behaviorally divergent portion of the population behaves in a random fashion as compared with the bulk of the population. This behavior is analogous to adding noise to code, at prearranged intervals, in order to mislead would-be interceptors of the message (Chance and Russell, 1959). It is these "randomized" displays that Chance has termed Protean.

For example, some of the mice of the genus *Peromyscus* are susceptible to audiogenic seizures. Similar seizures in these mice may be precipitated when they are under attack in a situation affording no opportunity for hiding. The seizure consists of violent undirected running, followed by either catalepsy or violent, often undirected, aggression. It has yet to be shown, of course, that such behavior is actually sufficiently disconcerting to a predator to render it useful (either to the individual under attack or to its species.) It does seem a reasonable assumption, however, that any increase in the complexity of the predator's stimulus field will accrue to the advantage of the prey. Viewed in this light, the susceptibility to audiogenic seizures, itself a polyethic trait, may well represent an example of a Protean display.

Roeder (1959) and Treat (1955), in describing the random movements of noctuid moths that have detected the presence of hunting bats, provide us with an instance of a Protean display that is manifested by all members of a population rather than only a fraction. These moths have receptors sensitive to the ultrasonic calls produced by free-flying bats and respond instantly to such a stimulus by altering their flight patterns. This relationship between the behavior of a vertebrate and an invertebrate raises intriguing questions about the manner in which two disparate types of nervous system can function to receive similar kinds of stimuli with roughly equal latencies (cf. Roeder, 1959).

Motivational levels

When all other variables affecting the adequacy of a mimic's protective devices have been considered, one must acknowledge changes in response tendencies of the predators. Both blue tits (*Parus caeruleus*) and great tits (*Parus major*) prefer large prey to small, other factors being equal, with the upper limit of the size of the acceptable prey being largely a function of bill size. As the broods of these tits increase in size, however, the size of the preferred prey increases, too (L. Tinbergen, 1960); thus the risk run by the prey is not merely a function of its own palatability, abundance, and crypticity. Tinbergen also suggests that the search image plays a role primarily at intermediate, and not at very high or very low, densities of the prey. It is thus clear that changes in the predator's response tendencies may play a crucial role in experiments dealing with mimicry and predation.

Such changes in motivation or response tendency may also occur in ontogeny. Vince (1960), for instance, has shown that in great tits the tendency to approach a nonrewarding stimulus varies with age. Systematic studies of the effect of age on the ability to discriminate test objects will be necessary before the crude experiments dealing with the effectiveness of mimicry can be validated.

Finally, it must be remarked that there is evidence (Logan, 1961) that suggests that a learned discrimination may be specific to a particular location. Rats that were trained to make a discrimination in one locus had to be retrained at a second one. Russian workers have provided yet more support for the view implied by these results, that learning must not be viewed as cut from one cloth: Ecological and evolutionary factors have shaped the capacities of each species separately. Some of the relevant Russian studies have been summarized by Kovach (1971). He writes (p. 19):

> Studies have shown that stimuli natural to the given organism in its ecological setting result in exceptionally quick conditioning. On the other hand, stimuli which are not normally encountered within the organism's normal ecological setting may require much longer time periods before conditioned responses are established. In fact, some ecologically strange stimuli cannot be used for conditioning at all. Thus, studies by Klimova (1958) have shown that the sound of rustling leaves or tactual stimulation of the upper part of the neck in wild

hares may serve as unconditioned stimuli in a very quick establishment of conditioned alterations of respiration rate. In domesticated rabbits, such conditioned responses cannot be established at all. Similarly (Slonim, 1961) conditioned reflexes can be quickly established in ducks (often after a single trial) when species-typical vocal calls or natural environmental sounds are used as conditioned stimuli. Conditioned reflex reactions to the vocal signals were found to be practically inextinguishable. Slonim (1967) argues that such rapid and persistent establishment of conditioned responses is possible because of the involvement of unconditioned reflex responses to species-typical vocal signals and natural environmental stimuli.

Uzhdavini has demonstrated in Slonim's (1967) laboratory that foxes (*Vulpes* sp.) rapidly develop conditioned reflexes in the process of digging a mouse (*Mus* sp.) from its hole. Foxes raised in captivity and having had no experience in hunting and capturing mice will learn to dig at an artificial hole after only two reinforced trials. Similarly conditioned change in respiration rate to vocalization of mice can be established in the fox by only one trial of consuming a mouse, and this response is practically inextinguishable (Goleva, 1955). Apparently in the fox, species-typical behavior of responding to the vocalization of mice and digging has very strong unconditioned reflex foundation and can be integrated into an effective behavioral chain by an extremely quick conditioning process (Slonim, 1967). On the other hand, studies by Voronin (1957) indicate that certain types of species-typical movements, such as specific tail movements in the dog (*Canus* sp.), cannot be used for conditioning with food as a reinforcement. These, and numerous other related evidences (see in Birivkov, 1960; Slonim, 1967) show that the ecological adequacy of stimulation plays an important role in learning.

Evidently, more attention to the importance of different environmental cues and their relation to sensory and perceptual capacities can well be given by behaviorially minded biologists.

In all of this discussion we have taken for granted the reality of mimicry, that is, that a vulnerable animal does attain some protection by resembling a distasteful or dangerous model. The behavioral problems involve defining the limiting conditions that must be met for mimicry to be effective. Many naturalists may resent our having begged the fundamental question: Is selective predation in fact the cause of mimetic resemblances? Just because it might theoretically be so, or has been shown to be so in specific cases, is not compelling. The doubters have a case: The possibility that predation is related to relative abundances rather than appearances cannot be dismissed, and the number of conclusive field studies of mimicry remain pitifully small (but as examples of those that have been done, see Cook et al., 1969 and Eisner et al., 1962).

TWO

How Are Food and Space Shared between Species?

It has become a routine exercise for ecologists to demonstrate that sympatric species do not share the same foods. Where the same items are consumed, interspecific differences are presumed to exist in the sizes of the objects taken or in the time of day when feeding occurs. All this information is just what would be predicted from the classic studies of competition, the exceptions being those equally predictable instances where food supplies are practically unlimited and exert no influence on population growth. Andrewartha and Birch (1954) apparently consider this to be the case with many Australian insects that are subject to periodic and irregular decimation by climatic extremes. Competition for food is not believed to occur since food supplies can be considered unlimited.

The situation of greatest biological interest, however, is the one where food *is* limited and interspecific competition for food is nonetheless avoided. The ultimate, or evolutionary, causes of this diversification of feeding habits have been examined by Volterra (1931), while a major series of experimental studies relating to the same problem has come from Park (1948) and his students. What the latter studies have shown, essentially, is that where diversification is possible, as in a heterogeneous environment, two potentially competing forms coexist, while in a homogeneous environment one will replace the other.

The dimensions along which the degree of homogeneity is measured may vary from case to case, of course. While in some cases the diversity depends on differences in size (Hutchinson, 1959), it may more often depend on differences in location or manner of feeding (Schoener, 1968). It may also be due to differences in the time of feeding (Cleach, 1967) or to a combination of these factors. Collias and Collias (1963) describe the selective feeding of ducklings (*Anas* sp.) of different species as being due to both difference in feeding methods and date of hatching. In the pages that follow we shall therefore have to examine the behavioral basis for a stable diversification of both feeding habits and feeding (or living) areas.

INTRODUCTION TO THE PROBLEM OF FOOD

The particular question to which we shall address ourselves is, How are specific feeding habits or food preferences maintained from one generation to the next? Captive animals are frequently capable of utilizing a wide range of unusual substances for food, narrow preferences as those of Jack Spratt being the exception rather than the rule. This is particularly true of birds, which are largely catholic in their tastes when placed in noncompetitive situations. How, then, are the naturally occurring differences between sympatric species maintained? In the paragraphs that follow we shall distinguish several different determinants. It should be clear, however, that these do not operate independently of one another. Distinctions made for expository purposes may be required even where they magnify slight differences.

Structural Determinants of Behavior

Many species have developed so specialized a feeding apparatus that their anatomy mechanically limits variability in their feeding behavior. Such limitation can occur to varying degrees. Among birds, for example, there are some species with bills so structured as to allow for a wide choice among food stuffs, while others, like the flamingo (*Phoenicopterus ruber*), are severely limited by the design of their bill.

Where the design of trophic appendages does not determine the nature of the food, there may, of course, exist other structures that play a rigidly determining role. The presence or absence of an enzyme system equipped to deal with particular substances (e.g., wax, as in the honeyguides—Friedman, 1955) or neural configurations that mediate "innate" preferences for particular shapes or colors are examples. Hess (1956), for instance, has demonstrated that naive (inexperienced) chicks may show differences in their responses to grains dyed different colors. The obvious point to be made is that, ultimately, there is no valid distinction to be made in behavior between structure and process, a point whose implications have been elegantly expounded by Whorf (1956). However, if by "structure" we limit ourselves to external appendages, we can meaningfully contrast species that are rigidly limited by their form to those that are less limited.

Both the towhee (*Pipilo erythrophthalmus*) and the catbird (*Dumetella carolinensis*) may feed upon the forest floor, seeking insects and their grubs. The towhee manages this task with considerable poise and efficiency. With quickly alternating movements of his feet, he scratches aside the leafy litter, spearing each insect as it comes to view. The catbird rather deliberately pokes about the litter, lifting leaves with his bill to examine the underlying substrate. As might be expected, catbirds do not obtain much of their food on the ground. Even on the island of Bermuda where towhees are absent or uncommon and where the catbird has become one of the three most abundant species, it has not modified its leisurely technique (Crowell, 1961; also, personal communication). The absence of competitors, according to Crowell, allows the three commonest species of Bermuda to attain densities considerably higher than these species attain on the mainland, even in the best of habitats. This fact notwithstanding, the catbird exploits the vacant microhabitats only by utilizing its conventional and characteristic feeding motions. Crowell has raised the provocative question of the result of an introduction of towhees to Bermuda. Being more efficient at ground feeding than catbirds, would they compel the latter to restrict the bulk of their feeding to other sites? Or are certain individuals of the catbird population, rather than all individuals, responsible for the expansion of catbirds into towhee habitat? In this event, only a proportion of the catbird population would be affected by the introduction of a competitor. The range of feeding movements of the entire population would remain the same, but the diversity would differ. This process can be illustrated as in Fig. 2–1. If the curve describing the range and proportion of feeding movements for the entire species (*a*) is composed of a mosaic of individual curves with their peaks along different points of the abscissa (*b*), then the total diversity of feeding responses of the species would be diminished upon the introduction of a competitor such as the towhee. On the other hand, if the situation depicted by (*c*) obtains, it is diversity of movements of *individual* birds that would be affected. This last instance would suggest considerable plasticity in the individual catbird's feeding repertoire, the former (*b*) a lesser degree of phenotypic but perhaps an enhanced degree of genotypic plasticity (cf. Chapter Three).

Since it has often been suggested that differences, such as in the catbird's and towhee's mode of feeding, can be attributed to differences in the musculature of the legs, it would seem possible to determine, at least in this case, whether a given range of feeding movements found in a large population is due to the plasticity of behavior of each individual or to a structural polymorphism characteristic of the entire species. An initial comparison between birds of the same species from Bermuda and North Carolina indicated that the insular individuals were slightly less stereotyped in their feeding activities and foliage preferences (Sheppard et al., 1968). The study of birds that have colonized islands by J. M. Diamond (1970) also suggests that continental birds are initially more stereotyped. The colonists may expand their habitat vertically rather quickly, but diet and foraging techniques are altered only after prolonged isolation. More will be said of this point in our discussion of character displacement in Chapter Three.

Finally, a word should be said about the possibility of structural relations (exclusive of the central nervous system itself) determining prenatal learning. For example, the characteristic pecking movements made by the chick (*Gallus* sp.) in hatching have frequently been alleged (Kuo, 1932) to have been established as the consequence of rhythmic contractions of the cervical musculature that represent forced oscillations triggered by the heartbeat. This explanation was originally offered as a counter to the view that complex sequences of acts were directly encoded in the central nervous system by the genetic material. Such nativistic views have been criticized on a

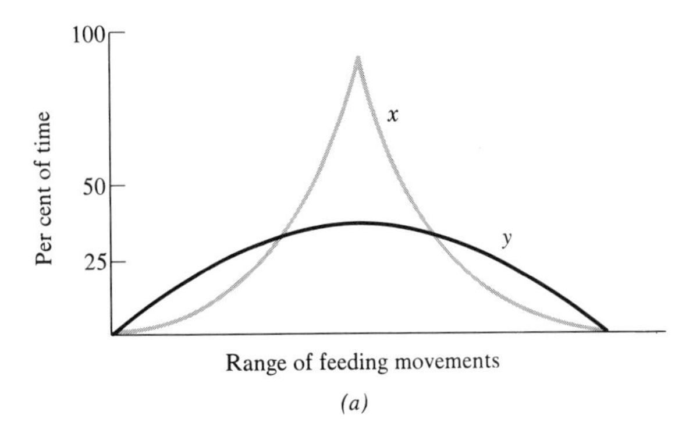

(*a*)

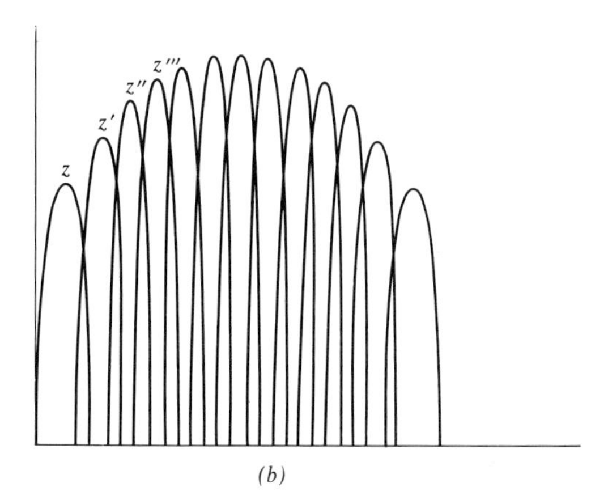

(*b*)

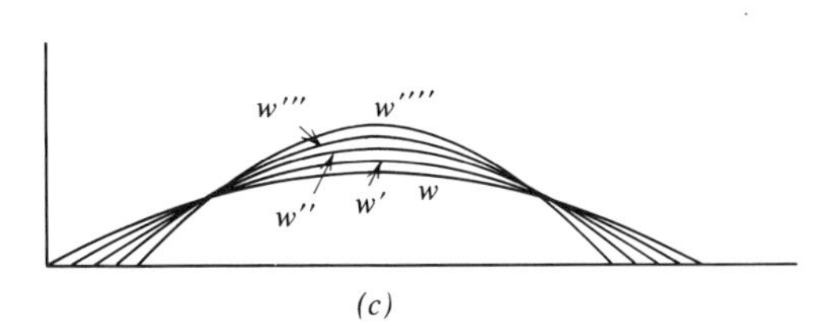

(*c*)

Fig. 2–1 (*a*) Curve y describes the proportion of time members of a species spend on different types of feeding activities. This is a species *without* competitors in the same habitat. Curve x relates to a species *with* competitors in the same habitat. (*b*) Curve x described in (*a*) could result from the addition of the curves of a group of specialists, each curve corresponding to the activity of one individual. These individuals are "masters-of-one-trade." (*c*) Curve x could also result from the addition of these curves, each one describing the activities of one individual. These individuals are "jacks-of-all-trades."

variety of grounds (see Klopfer and Hailman, 1967, and Marler and Hamilton, 1966). Kuo was also critical of the more extreme forms of environmentalism, but this aspect of his writing has been less often noted (but see Gottlieb, 1970).

The notion that acts represent the direct expression of a gene is based on a number of misconceptions concerning the nature of gene action. Even so simple and paradigmatic an act as the pecking of a chick is far from a unitary affair. Seemingly identical pecks may involve wholly different muscles on each repetition. Must we postulate a gene for every possible nerve or muscle fiber that contributes to a given response? (See the final section of Chapter Four for further details.)

Behavior is not a noun, defined and determined by a discrete locus on a DNA molecule. It is a process that derives from a series of interactions, some stochastic, some perhaps deterministic, that at times can achieve a certain level of predictability and stereotype. At some point this degree of inevitability and sameness becomes so great that we speak of "an instinct." But let us not neglect the fact that acts, or behavior, are sequences of movements and perceptions and are best described in terms of latencies, frequencies, durations, and intensities. They must be analyzed with a view to unraveling a skein of interactions that tie together particular stimuli and particular responses.

Is it equally illusory to explain behavior without reference to the nervous system and its related structures? Gottlieb (1970) provides a concise summary of the opposing views, which represents a shift from the nature–nurture dichotomy of the 1930s:

> The single most important issue dividing the two theoretical camps is the role of endogenous and exogenous stimulation in behavioral, neuroanatomical, and musculoskeletal development. At the level of functional anatomy, there are a few recent prenatal experiments that suggest that relevant sensory and musculoskeletal stimulation may be essential to the initiation and maintenance of normal (typically observed) maturational changes. These and other experiments with neonates would seem to indicate a particularly significant shift in our conception of the structure–function relationship from a unidirectional one (structure → function) to a bidirectional one (structure ↔ function). This shift is consonant with a probabilistic conception of epigenesis and raises an important

> question for future resolution at the prenatal level, namely, the degree to which stimulation or activity merely fosters or enhances development and the degree to which (or areas or stages in which) stimulation or activity channels future development. [P. 134.]

Examples of forms of behavior established along lines similar to those proposed by Kuo for the chick (*Gallus domesticus*) now abound. However, Oppenheim (1970) found that movements of embryonic chicks and ducks were not influenced by manipulating the limb positions of the embryos. In a contrary case (man), Margaret Mead (personal communication) has suggested that the culturally distinct—and culturally determined—stances of pregnant and recently parturient women may directly affect the postural stances of the young infant. Environmental factors may accelerate or retard developmental rates, thereby altering behavioral responses. In the case of domestic ducks (*Anas platyrhynchos*), Gottlieb and Klopfer (1962) have shown that the tendency for the newly hatched young to learn the auditory or the visual aspects of an *imprinting* model is a function of their developmental age. Early hatchers tend to be auditory learners, focusing on the acoustic qualities of a complex model that they have learned to recognize. The later hatchers, however, focus on the visual qualities of the model. Thus such factors as CO_2 levels and relative humidity, which affect the time of emergence from the shell, can also determine the dominant sensory modality during imprinting (at least in the Pekin duck).

Whether, and how, stimuli received by the mammalian fetus or avian embryo affect later development and behavior will have to be established anew in every case. Situations such as those described by Kuo (1932) will require mechanical interference with embryonic movements for their elucidation (along the lines attempted by Oppenheim, 1970). Where changes in thresholds are involved, as may be the case in the duck situation described above, the problems are of greater subtlety. In those cases where sensory or perceptual habituation is involved, changes in the stimulus field can be made directly. The schooling behavior of fish may depend on their aligning themselves tail to eye. During embryogenesis, the fish are curled within the fetal membranes with tail opposite eye. Perhaps, as Lorenz (personal communication) has jocularly suggested, it is the stimulation afforded by the slowly undulating tail during development that provides the basis for the responses that lead to schooling in the fry. (Also see Oppenheim, 1966; Hamburger, 1963; and Gottlieb, 1970, for further detailed discussions.)

The Interplay of Structure and Learning

We have suggested that cases where gross structure rigidly limits behavior are relatively uncommon. Much more abundant are those instances where anatomical features affect the ease or efficiency with which different types of prey can be handled. A cardinal (*Richmondena cardinalis*), for example, can rapidly devour sunflower seeds, splitting the seeds along their edges and deftly maneuvering the kernels into its mouth. A chickadee (*Parus atricapillus*), if the seed be large, must hammer it into fragments before it can extract the pieces of kernel, a considerably slower process. Given millet, it is to be expected that the chickadee would operate the more efficiently of the two. Kear (1960) conducted an analysis of the role of experience and bill shape in the development of food preferences in various passerine birds. Largely confirmed by her experiment is her assumption that small differences in the ease of handling certain foods would lead to a learned preference for the more easily manipulated item. Other conditions being equal, according to her findings, a bird will develop a perference for the largest seed it can efficiently manage, efficiency being measured by the weight of kernel extractable in a given length of time. Where an increase in the size of the seed taken leads to a drop in efficiency, a preference for a smaller seed develops. Hespenheide (1966) and Newton (1967) made similar findings.

Lehrman (1956) has provided an even more ingenious model to explain the development of parental feeding behavior in the ringdove (*Streptopelia risoria*). The young of this species are, at first, fed a regurgitated secretion from the lining of their parents' crops, the so-called "pigeon milk." Lehrman was able to show that inexperienced pigeons never provided their young with this food until after the young had repeatedly poked their bills about, into, and around the breast and mouth of the parent. Experienced parents, on the other hand, frequently initiated feeding prior to receiving any tactile stimulation from their young. Upon learning that vomiting could be induced by compounds such as yohimbine hydrochloride and thus could serve as the terminal response in a conditional response chain, Lehrman erected the following hypothses: The presence of prolactin in the bloodstream (induced by earlier events occurring during the breeding cycle) causes an engorgement of the epithelial tissue of the crop. The tension this engorgement creates is relieved, temporarily, by regurgitation. Initially, in the inexperienced bird, a tactile stimulus is required for the elicitation of the vomiting response. The relief that is obtained serves as reenforcement, and thus, on future occasions, the sight of a squab with bill raised is sufficient to elicit the regurgitation and feeding response. Insofar as the young are concerned, the reception of food from an adult would reenforce any tendencies to direct the begging response in the appropriate fashion. Thus experience can modify the consequences of structural restrictions (including endocrinological and reflexological ones).

The limit to the variety of outcomes of processes such as those suggested by Lehrman is unknown. Similar models have been invoked to explain such traits as aggression or obesity in humans. For example, firstborn or only children (in the United States) are particularly likely to be obese. This was explained by Schacter (1968) by considering the response of a newborn infant to all those events that provide sensations of discomfort—it itches, it is cold, wet, hungry. To all these, and sometimes spontaneously, it responds with a wail. Its unskilled mother responds to all wails as if they signaled hunger. The child may be warmed, dried, succored, and so forth, but only after it has first been fed. Feeding becomes an intervening signal: physical distress—feeding—then relief. These infants, then, as adults, are programmed to respond to all conditions of distress (including anxiety) by feeding, and hence their obese condition. In them there has presumably been established a deviant, though stable and predictable, set of stimulus–response contingencies. Presumably many (any?) responses could be conditioned to many (any?) emotional states, just as we can be conditioned to subjectively either value or dislike particular emotions (i.e., a particular emotion can presumably function as a positive or as a negative reenforcer). Where variations of this kind do not occur, the contingencies have apparently been fixed at an earlier stage in development—a reasonable evolutionary stratagem for responses that need not be variable and that occur to stimuli that do not vary.

An example of a somewhat different kind has been recorded by Hailman (1961), who had been struck by the incongruity of chicks of the laughing gull (*Larus atricilla*) pecking at contrasting spots. In most other species of gulls the adults wear a contrasting spot on their bill. Thus the spot-oriented pecking of the young brings them into close contact with the source of their sustenance. Even among domestic chicks (*Gallus gallus*) this behavior is adaptive, assuring that the young will eventually discover edible particles. The laughing gull chicks, however, do not peck at food on the ground in the manner of domestic chicks, nor are the adults endowed with any specific markings. The bills of the chicks themselves, however, do have contrasting tips. Hailman describes the role of this bill pattern as follows:

> The three naive chicks which were closely observed learned to discriminate food in the following manners: one chick appeared to learn to peck at food through trial-and-error exploratory pecking. However, each of the other two first received food by pecking at the bill-tip of an experienced companion, which at the same time was pecking at fish. In identical fashion, both of these previously naive chicks then pecked quickly three times in succession at pieces of fish. One bird missed food on the first peck, but received some on the next two pecks; the other bird received food on the first, missed on the second, and was again rewarded on the third peck.
>
> Thus, it appears as if two of the three inexperienced chicks initially found food by pecking at the bill-tip of an experienced sibling and then immediately established a discriminatory response. This constitutes a process of social interaction in which the presence of an experienced individual facilitates the learning process in an inexperienced individual, and is somewhat similar to the processes included under "empathic learning" by Klopfer (1959a). [P. 246, 1961.]

Given the relatively greater plasticity in the behavior of most mammals, it would appear likely that this role of individual experience would assume a correspondingly greater role as we ascend in phylogeny.

The Role of Tradition Learning

From an influence of the individual's previous experiences on his choice of foods, it is but a short step to the situation in which ancestral experiences are collectively transmitted to succeeding generations. This type of learning, where parental example replaces individual experience, is obviously a primary characteristic of human behavior. There are, however, strong reasons for believing this to be a situation not unique to *Homo sapiens.* First, it has been shown (Chapter One) that a human level of behavior need not be attained for there to be secondary conditioning, a type of learning in which it is the responses of a congener that become the relevant stimulus for learning. Second, many organisms have developed a degree of social organization of a highly stable and elaborate nature. A herd of deer (*Cervus* sp.) (Darling, 1937) and a colony of bees are two of the more obvious examples of groups in which the occurrence of secondary conditioning would be promoted. Finally, in some forms there is an extended period of filial dependence on the parents [up to two years in the elk or moose (*Alces* sp.) (Altman, 1958)].

These three characteristics would seem, a priori, to be all that is required for the transmission of certain conventions from one generation to the next. In point of fact, the psychological literature is rich in examples of observational learning, principally among primates, but also among such diverse forms as birds, cats, and prairie dogs (*Cynomys* sp.). Miyadi (1959) has recently chronicled the spread of a peculiar habit among a colony of Japanese monkey (*Rhesus* sp.). A particular individual undertook regularly to wash the sweet potatoes given it as food. Its mother and those of its peers who most frequently played with it were the first to copy the action, the behavior gradually radiating outward until most of the animals of the colony had adopted it. The older the monkey or the higher its position in the dominance hierarchy, the longer it took until it, too, had adopted the trait. One can safely predict that the sweet-potato-washing habit will continue for some time to differentiate this colony from others, just as different colonies have developed different food preferences.

Among oyster catchers (*Haematopus* sp.), there are some that prepare the mussels on which they feed by stabbing into them and others that hammer them. The procedure that is used is apparently a matter of family tradition (Norton-Griffiths, 1966).

The argument that culture, or the nongenetic transmission of traditional modes of behavior, is dependent on the acquisition of language can, in view of the preceding facts, scarcely be supported. Broadbent (1958), indeed, has argued that the principal function of language is to allow transfer of large quantities of information rapidly. The development of linguistic abilities is thus considered to be a function of a nervous system large and complex enough to deal with many items simultaneously and is not the sine qua non for the use of symbols. A particularly striking example of this phenomenon comes from the work of Cooper (1956), who has shown how a eumenid wasp, whose eggs may be deposited serially within hollow bamboo tubes, "directs" the larvae to the exit. This information is imparted by the partitions the female constructs between each two eggs—direction of curvature or the roughness of the partition represent the cues to which the larvae must respond. Redundancy in communications can thus be seen to arise rather low in the phylogenic scale. Similarly, the transmission of information about food and habitat need not await the development of linguistic abilities.

The Nature and Effects of Imprinting

A special type of learning should be mentioned that may closely mimic the effects of other kinds of tradition learning. This is a relatively rapid learning of certain general attributes of a stimulus object, and it appears to be limited to a more or less fixed developmental stage. It is also relatively independent of the overt reenforcement usually required for other types of learning. In fact, it shows striking similarities to latent learning, especially since the responses to the stimulus object may not appear until long after the original exposure. The term that has been applied to this form of learning is *imprinting*. Long before imprinting had been formally recognized as a somewhat anomalous form of associative conditioning, it had been shown (Thorpe and Jones, 1937) that the host preference of some parasitic insects could be determined by the olfactory stimuli to which the individuals were exposed as larvae (also see Manning, 1966).

As this phenomenon was observed more frequently, it became characterized in the following way:

1. The exposure of a young praecox (such as ducklings, *Anas* sp.) to a particular model could lead to a subsequent preference for that model only if the exposure occurred during a specific and limited period after hatching, the so-called critical period.
2. A preference, once established by exposure during the critical period, was irreversibly established.
3. The strength of the praecox's preference for the model in-

creased linearly with the amount of energy it expended during its initial exposure; that is, a duckling that was drawn behind a model in a small, wheeled cart would show a weaker preference than one that struggled over hurdles while following the model. (It must be noted that praecocial young of many birds tend to follow any moving object to which they are exposed during the critical period, thereby assuring "imprinting" onto the object that originally elicits their following response.)

Other supposedly characteristic features of this phenomenon have been summarized by Hess (1959), Sluckin (1965), and P. P. G. Bateson (1966), though the significance of some of these (for example, differential effects of various drugs) is far from clear.

Since we wish to stress the fact that imprinting does not reflect a unique form of learning, it is worthwhile to consider the criteria by which it has been defined (also, cf. Thorpe, 1956). First it must be noted that the occurrence of the critical period, in terms of age measured from the time of emergence from the shell, differs from one worker to another, even for the same species. At one time it was felt that this difference could be due to differences in genotype, particularly among inbred strains. It was also suggested (Klopfer and Fabricius, personal correspondence) that the difference lay in the manner of incubation. Oil-heated incubators may tend to have higher CO_2 levels within them, which would delay hatching. However, it was noted that even among synchronously incubated eggs taken from a single clutch, up to 60 hours might elapse between the hatching of the first and last egg (the usual range was 12 to 16 hours) (Klopfer, 1959b). If one takes 12 to 16 hours, posthatch age, as the critical period for Pekin ducks (*Anas platyrhynchos*), it is clear that two individuals tested within this period might differ in their developmental age by more than two days. This observation led Gottlieb (1961b) to plot the percentage of duckling that could be imprinted against age, taken in terms of hours since both the onset of incubation (developmental age) and the emergence from the shell (posthatch age). A much more sharply defined critical period was evident with the former measure. What this study suggests is that experiential factors are of secondary importance to developmental factors in the determination of the critical period. It also suggests that some young of any clutch may hatch at an age already beyond the limits of the critical period. Where this is the case, bounds are inevitably set on the effectiveness of the imprinting mechanism for assuring the learning of certain characteristics.

The second criterion, that of the stability or irreversibility of the preference for the imprinted surrogate, is based largely on anecdotal accounts. When a gosling (*Anser* sp.) follows a man for a brief period during its critical period and, upon reaching sexual maturity, directs its courtship toward men, one can speak of this being an imprinted preference only if no rewarding situations associated with men intervened. Experimental determination of the nature of an imprinted "preference" has generally depended on one of three operations. First, it has been based on observations of courtship and mating behavior. These observations assume that sexual preferences are the result of imprinting: The preference reveals the character of the imprinted pattern. Indeed, the original connotation of the term imprinting, when introduced by Lorenz (1935), did refer to sexual preferences, ignoring other kinds of preferences, as for certain foods or habitats, that may be similarly established. An economic drawback of examining sexual preferences in order to discern the character of the imprinted pattern is the requirement that the subjects be reared for a relatively long time, either until sexual maturity is attained or at least until overt sexual behavior can be elicited by the administration of exogenous hormones (e.g., Andrew, 1964, and Schutz, 1965). A second procedure, based on the assumption that the presence of the imprinted object is more rewarding than its absence, has been to use the momentary appearance of the imprinted pattern to reenforce an act in an instrumental conditioning paradigm, such as key pecking (e.g., Hoffman et al., 1966). Presumably, objects onto which no imprinting has occurred will fail to serve as reinforcers. However, this, too, is a tedious procedure and, as Hoffman et al. (ibid.) have noted, raises a number of puzzling questions on the nature of reenforcement. Finally, and most commonly, an approach to and "following" of the imprinting object has been used as the measure of an imprinted preference. This has been particularly true in studies with neonatal domestic chicks and ducklings, which tend to follow objects moved before them. If the subject is imprinted, the following response is believed to become preferentially directed to the original test object. However, the work of Gottlieb (1961a) has also indicated that the preference of a duckling for the surrogate it followed during the critical period is not absolute. Given a choice between two surrogates, the duckling will follow the one originally presented. In an absence of a choice, it may accept another. This introduces an element of plasticity into the animal's behavior, which it may exploit to good advantage, as we shall see.

Other puzzling discordancies appeared. For instance, it was possible to group a series of pairs of models into three categories with respect to whether or not they were discriminated from one another following imprinting. However, these categories broke down if the models were stationary rather than moving; i.e., models that were discriminable under conditions of movement ceased to be so when not moving.

To resolve the discordant results with moving and stationary models, their discriminability was studied when they were presented in a conventional learning paradigm (Klopfer, 1969a and b). The question here was to determine whether discriminability was simply related to the subject's sensory ability or depended somewhat on the imprinting situation. It was then shown that models that were treated as equivalent (nondiscriminable) in the imprinting situation could be readily discriminated in the learning paradigm (also see Zolman, 1969). The nature of the reenforcing conditions (whether it was heat provided to a cold subject or water to a dehydrated one) proved important, too (also noted by Garcia et al., 1968). In any event, it appeared that the results in the imprinting tests could not simply be attributed to peculiarities in the subjects' peripheral sensory capacities but was in some way related to the context in which the models appeared.

Regarding the third criterion proposed by Hess (1959), one can only suggest that it remains for his work on the "law of effort" to be replicated with species other than domestic mallards (*Anas boschas*). "Effort" is unlikely to be involved in the responses of wood ducklings (*Aix sponsa*), which can be auditorily imprinted. So the "law" is, at most, of doubtful generality. Clearly, more purely descriptive studies of imprinting are well justified.

It is possible that the return of birds of many different species to the breeding grounds of their parents may represent imprinting onto a particular locality. This process, then, can also account for the maintenance of food or habitat preferences, as well as provide an explanation for the explosive rapidity with which a species may come to occupy a new habitat. That is, if an adult rears her young on a "new" food or in a "new" environment, providing that the species in question is susceptible to imprinting, her young, irrespective of genotype, will all tend to perpetuate the new habit (Wecker, 1969; Hovanitz and Chang, 1963).

The susceptibility to imprinting is not equal among all species and, in fact, seems definitely linked with particular developmental rates. The susceptibility is greatest in species whose young show a high degree of locomotor precocity and, at the same time, normally maintain social contacts with their parents and siblings for a period of weeks or months. In this category fall the Anatidae (Hess, 1959) and many ungulates (Collias, 1956; Altman, 1960). Olfactory imprinting in insects, if it does prove to be a comparable phenomenon, represents somewhat of an exception. In addition, if evidence can be adduced linking the critical period for Freud's infantile fixations in humans with the critical period for imprinting, then, of course, the presumed relation between motor precocity and imprintability must break down altogether.

At the present time, then, we are inclined to regard imprinting as a form of behavior highly adapted to the needs of creatures with rapidly maturing motor abilities. Ewer (1956) has suggested that it may represent a stage in the evolution of an endogenous act (whatever that is) from one that was learned through repeated experience. Functionally, it does indeed serve to assure predictability of response patterns while preserving flexibility in their target.

INTRODUCTION TO THE PROBLEM OF SPACE

In a good many instances, space may represent an important and immediately limiting element in the expansion of a population. Among colonies of cliff-dwelling seabirds, for example, it is not unusual to find that every inch of available nesting space is utilized. However, the limiting effects of space may be manifested in other ways than through directly limiting the number of nest sites. Gibb (1960) has shown that great tits (*Parus major*) and goldcrests (*Regulus regulus*) require about one insect in the phenomenally brief period of every 2.5 seconds for approximately 90 percent of each day in order to maintain themselves during the winter. Territorial behavior in the autumn acts as a proximal regulating factor, keeping any local population at sufficiently low levels of density to avoid the likelihood of a major decline from food shortage in the winter. Thus food shortage serves as the selective force maintaining spatial separation. Space limitations may also express themselves in behavioral disturbances, as the disruption of courtship that occurs when *Drosophila* are too densely crowded, or the increase in adrenal size and concomitant increase in probability of death in dense aggregations of rodents and rabbits (family Leporidae). In short, space may be considered as restricting populations directly, by limiting nest or breeding sites or feeding territories, or indirectly, as the result of physiological responses to crowding. Indeed, Christian (1970) has argued that more rapid evolution of some taxa is due to their intolerance to crowding and subsequent dispersal to new habitats and new selective pressures.

All this information implies that there is competition for space, both on an inter- and intraspecific level. Thus one can infer that natural selection will lead to the appearance of devices or stratagems to mitigate the effects of the competition. The result expected is a diversification of the different species' requirements for space and a regulation of the requirements of individuals of the same species. We shall consider these two aspects of the problem separately.

Let us first consider how space might be apportioned. We could make divisions with horizontal coordinates, producing layers at different altitudes to each of which a particular species is relegated. This, in fact, is the situation among littoral and marine species, the horizontal stratification being immediately due to differences in tolerances for salt desiccation or hydrostatic pressure. Another possibility would be to divide a space with vertical coordinates, restricting some species to one area and others to another. This situation obtains where the distribution of a species is determined by a geologic accident, as when the Panamanian Isthmus was submerged and separated some of the Southern and Northern American rodents. Finally, if our space is at all heterogeneous, we might divide it into units of a lesser degree of homogeneity, that is, into microhabitats. Thus the portion of the space atop the forest canopy might be reserved for one group, that close to the trunks of trees to another, and that beneath the bark to yet another. Summarizing, we may distinguish (1) species whose ranges are determined by accidents of geology and history (their distribution corresponding to a division with vertical coordinates), (2) those that are limited to a particular horizontal zone by structural or physiological adaptations (division with horizontal coordinates), and (3) those that select a specific microhabitat irrespective of locus in space, either because of (a) a morphological constraint that so limits them or (b) social influences, representing a form of tradition learning.

The Role of Historical Accidents

The limitation of marsupials to particular regions serves as one example among many of how geographic isolation can restrict the distribution of a group of animals. Disjunctions or discontinuities such as those posed by the Isthmus of Panama, the Bering Sea, or, on a less gross level, the Colorado River or Mojave Desert have proved to be barriers to an entire spectrum of forms. These barriers effectively apportion space between species. Recent detailed summaries of biogeographic knowledge obviate the need for a comprehensive summary of case histories (Darlington, 1957; Lindroth, 1957). Periodically a barrier may be surmounted, either as the consequence of changes in land or sea level or as the result of rafting or deliberate introduction. The explosive redistribution of the indigenous species that may result, including the extinction of some, is ample proof that their prior coexistence was solely a function of the happy accident of geographic separation rather than because of a diversification in habits (cf. Elton, 1958).

The Role of Morphological Adaptations

MacArthur (1961), arguing from the premise that a jack-of-all-trades is master of none, suggests that the microhabitat of bird species may be defined by a particular pattern in foliage density. After determining the foliage density at different heights and then preparing detailed censuses in a variety of habitats, MacArthur found that he could predict the presence or absence of particular avian species on the basis of the foliage density profiles alone. Of course, every good bird watcher accomplishes much the same, though the process of recognizing the appropriate habitat or microhabitat is largely intuitive and not susceptible to analysis. It is the merit of MacArthur's study that he shows how one can discriminate between essential and inessential cues in ones' predictions as to the occupants. Thus a knowledge of the floristic makeup of a community does not add to the accuracy of predictions made purely on the basis of foliage density measurements. What this study suggests is that space is divided horizontally, with the foliage density of any given layer and that of adjacent layers determining the presence or absence of any species. Within the

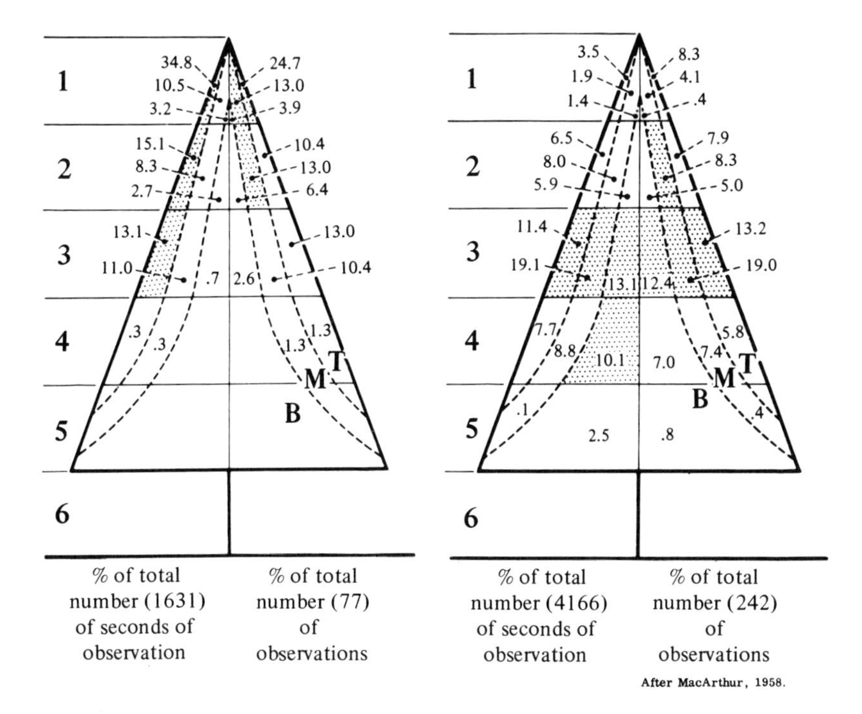

Fig. 2–2 Preferences of two species of *Dendroica* warbler are depicted in these schematic diagrams of spruce trees. The portions of the trees most favored by each species have been shaded until 50 percent of the time (or observations) has been accounted for. One species, it will be noted, is predominately in the outer edge of the upper three layers of the spruces; the other moves uniformly through the third layer. (After MacArthur, 1958.)

range of densities suitable to it, an individual might then be relatively unspecialized in the nature of his food or nesting site, competition having been reduced by means of a complex stratification (and note MacArthur and Pianka, 1966).

Whether the factors that determine the suitability of a given layer for a given species can be considered "morphological adaptations" remains open to experiment. That warblers whose diversified habits are indicated by Fig. 2–2 differ significantly in musculature or anatomy has never been demonstrated. The differences may lie on a perceptual level (cf. Chapter Three). At least in some species, anatomic differences can reasonably be claimed to exert a causal influence. The smaller body size of the blue tit (*Parus caeruleus*) as compared with the great tit (*Parus major*) allows the former to feed from the extremities of twigs that are barred to the latter. Similarly, the shorter-billed downy woodpecker (*Dendrocopus pubescens*) can deal with smaller shrubs than the much more heavily weaponed hairy woodpecker (*Dendrocopus villosus*) (William A. James, personal communication). As the different feeding stations of these four species are correlated with different foliage height densities, they provide an example of a physical character determining the stratum occupied.

Among mammals, foliage density must be of far less importance. Nonetheless, the structural differences that provide for spatial separation may be equally subtle. The three species of *Peromyscus* studied by McCabe and Blanchard (1950), for example, differed in the fine structure of their feet and toes, differences that affected their ability to climb or to penetrate loose duff and that can therefore be presumed to play a role in keeping them in separate areas. (With regard to habitat selection in amphibians and reptiles, note Sexton et al., 1964.)

Finally, limits to northward or altitudinal expansion of tropical or temperate zone forms may be due to factors requiring the attention of physiologists interested in problems of energy balance and temperature regulation. Certainly, differences in cold or heat tolerance can provide for the grosser sorts of spatial distribution. As for the experimental problems that these issues present, they are similar to those considered in the discussion of structural determinants of feeding habits, and we need add no more here.

The Role of Early Experience

Many references for a particular habitat or area have not been attributable to some physical necessity for restriction to the habitat or area in question. There seems to be no obvious physical necessity, for example, for salmon (*Salmo* sp.) to return to the stream of their youth, yet the olfactory experiences of the fry condition them to return, upon maturity, to the same stream, though other streams might serve them no less well (Hasler, 1956). Tits (*Parus*) can nest successfully in a variety of woods, though they show an overwhelming preference for particular kinds of woods. That psychological factors are very much involved in habitat selection is thus not a new observation (cf. Lack and Venables, 1939, and Miller, 1942), yet we are almost wholly without information as to the nature of these factors. Why are some species flexible in their choice of habitat and others less so? Under what conditions can one expect habitat imprinting to occur, in the sense that the first habitat experienced is inevitably preferred? What is the role of subsequent experience in a given habitat in developing specific preferences or aversions? What aspects of a habitat do particular species perceive and respond to? Does a tit recognize individual species of trees or the gross configuration of the forest? Problems of perception, discrimination, and learning are all involved, and as yet scant effort has been made to apply what is known of these phenomena to the problems of habitat selection.

A number of studies have been directed to the question of the nature of the stimuli perceived by birds in their choice of the appropriate habitat. It is possible to demonstrate, for example, that chipping sparrows (*Spizella passerina*) of either sex, trapped as adults in the North Carolina Piedmont will generally prefer to spend their time among pine foliage (loglolly pine, *Pinus taeda*) rather than in oak (several species of white, *Quercus alba,* and red, *Quercus borealis,* oaks). This is the case even where the perch opportunities are made absolutely identical by tying foliage to symmetrically arrayed metal rods. Feed is provided on the ground between the two rows of foliage so that no food reenforcement is associated with either plant species. A light gradient may be established along the long axis of the pine and oak, demonstrating that, within the intensities used, the amount of light penetrating the foliage is not important to the birds. By backlighting the foliage, shadows can be reduced, again without effect on the chipping sparrows. Hand-reared birds (from third to fifth day after hatching) behave similarly, although completely deprived of the sight of all vegetation prior to the experiment. White-throated sparrows (*Zonotrichia albicollis*), however, seem not to discriminate between these two foliage types, oak and pine, habitually selecting the darker portion of the light gradient. For these birds, shadow patterns or light intensity seem to play a larger role in habitat selection than leaf shape or color. A tentative hypothesis dealing with the basis for such specific differences in the nature of the relevant cues will be advanced in Chapter Three. For now it remains sufficient to stress the paucity of information that bears on this question despite the development of techniques appropriate to its study.

Loehrl (1959) has evidence suggesting that the learning involved in habitat selection may be limited in its occurrence to specific developmental periods (as with imprinting). His flycatchers (*Ficedula albicollis*) tended to return (in subsequent seasons) to the locale where they were reared, provided they had been free to fly about within it for two weeks prior to the completion of the juvenile molt. Those birds released after completion of the molt failed to return. It is possible that food supplies had diminished sufficiently in these cases to make the release locale unfavorable; that is, those birds held until completion of the molt were not released until late August when the southward migrations had already begun. Thus it is not known whether we are here dealing with a genuine case of habitat imprinting or a learning to recognize a habitat that exercises a reenforcing effect. Comparable problems have been described for amphipods by Meadows (1967).

INTRASPECIFIC DISTRIBUTIONS AND ABUNDANCES

Competition for a limited amount of space may occur within one species as readily as between different species. This competition is no less true of food or any other limiting factor in the environment. However, once the problem of how space is to be shared has been solved, the remaining questions may often be found to have been answered as well.

We may distinguish two aspects of this problem. First, we may ask what determines the spatial distribution or dispersal pattern of the species, and then we may inquire into the nature of the limitations on numbers within a given area. What these two forms of our inquiry assume is that the spatial pattern exhibited by a group of animals is independent of the absolute number occupying a given volume. There are doubtless animals for which this assumption can be shown not to hold, or to hold only within very broad limits. Artificially increasing the population density of such forms, as, for instance, the red deer (*Cervus elaphus*) (Darling, 1937), may disrupt the social forces that determine spatial patterns. Yet within natural populations of the red deer in Scotland, where one deer may occupy from 30 to 100 acres, social organization and thus spatial distribution remain constant over a wide range of densities. In other cases a rise in numbers may lead to a migration into marginal habitats (Pirowski, 1967, and Brown, 1963 and 1964).

What Determines the Spatial Pattern?

A limited piece of land may be apportioned in one of two ways. The simplest method is to provide each individual with a discrete and exclusive chunk. The decision as to which individual shall possess any given section may be based on arrival time, as in the case of many birds where territorial priority accrues to the first returnees from the winter haunts. Alternatively, it may depend on the individual's rank in a dominance hierarchy, as is the case with domestic fowl dividing up available perches in their roost.

Where the spatial organization is a function of the possession of the individual territories, the mean distance between individuals will remain reasonably constant. In many of the colonially nesting gulls (*Larus* sp.) and terns (*Sterna* sp.), for instance, this distance is almost precisely equal to the diameter of the circle within which one bird can peck another without having to leave its nest (N. Tinbergen, 1953). In various passerines the determinants of territory size, and hence spatial organization, are more subtle. The studies of Beer et al. (1956) on territory sizes on islands show that the minimum size may be grossly influenced by topographic features. However, it is equally clear that territories of many species are rendered incompressible by psychological forces. The implications of territoriality for the control of density will be dealt with below. The point to be made here is that a given spatial pattern results from territorial behavior and that this is relatively independent of the abundance of the species. Hensley and Cope (1951), for example, in their studies of the effect of avian predators on the spruce budworm (*Choristoneura fumiferana*), attempted to eliminate all birds from a particular woodland plot by intensive shooting. As quickly as a territorial pair announced its presence, the birds were shot, yet each day the territories were found to be occupied anew. Clearly, a substantial nonterritorial population must have been skulking about in the undergrowth, birds that had no posts in the pattern produced by the breeding population but that could quickly move into any that were vacated. This event implies that the territory size may remain constant even in the face of a rising density of individuals without territory, which, in turn, signifies that the breeding population possesses a spatial pattern of considerable stability.

What, however, determines the size and shape of a territory? In part these factors depend on the particular kind of territory that is involved. According to some preliminary studies on Manteidae (Jack Izower, personal communication), there are at least some species for which the spacing is an immediate function of food supply. When mantids are deprived of food, they become cannibalistic and intolerant of the proximity of a conspecific. Only when sated are they no longer spaced widely. In most colonially nesting marine birds, as indicated before, the territory serves merely as a locus for a nest and helps to assure the constancy of parental care upon the hatching of the young. Under these circumstances the territory becomes no larger than the largest area the inhabitant can easily defend. With many passerine species, however, as well as with mammals, the territory also represents the area within which food is collected. Such a feeding territory must, of necessity, be considerably larger. Those held by predatory and larger species are in fact larger than those of graminivorous or smaller species (Schoener, 1968). Other specialized forms of territoriality may exist as well (symposium edited by Hinde, 1956; Klopfer, 1969a and b, 1972), such as the display grounds of bower birds (Ptilonorhynchidae) (Marshall, 1954), but it is perhaps the more extensive feeding territories that require the closest scrutiny.

If territorial species are able successfully to rear more young than nonterritorial forms, the former will inevitably come to replace the latter. The development of territoriality, and of an optimum territory size, will then be a result of natural selection. Individuals with territories too small to provide life's necessities will be less likely to succeed in reproduction than those with larger territories, while those with excessive territories will find so much of their time spent in territorial defense that they also will be at a relative disadvantage. Of course, the maximum size of a territory might be expected to increase indefinitely as population density, and the need for defense, decreases. This does not seem to be the case, although in some species there is an expansion with the removal of neighbors (Watson, 1967). Whether territorial species learn that a territory of a particular size is more conducive to efficient food collecting than either one smaller or larger or whether topographic or floral discontinuities provide "natural" boundaries is not known. The study of the development of territorial behavior among naive organisms of several classes would appear to pose many intriguing prospects. These are rendered all the more difficult, however, by the fact that in some cases the size, shape, and location of the territory may undergo seasonal changes (Wiens, 1966, 1969).

Sometimes individuals of a species may be territorial in one region and nonterritorial in another. This is the case with the yellow-faced grassquit (*Tiaris olivacea*) (Pulliam et al., 1971). It is communal and nonaggressive on the Central American mainland and aggressively territorial on the nearby island of Jamaica. The explanation advanced for this difference is based on a model that deals with the evolution of altruism (see Chapter Six), but it can be summarized: An aggressive territory holder can decrease the fitness of a nonaggressive bird by excluding it from optimal habitats. The decrease in fitness of the nonaggressive bird is greater than the increase in fitness of the aggressor. However, the territorial bird does lose some of the advantages of social behavior (whatever they are) and must spend considerable time defending his territory, time that might otherwise be applied toward maintenance and reproduction. The amount of time that the average aggressive individual spends defending his territory must necessarily increase as the proportion of the bird population that is territorial increases.

Suppose that aggressive, territorial, individuals do have a lower reproductive capacity than the nonaggressive individuals would have in the absence of the former. Territorial behavior could then not evolve unless the populations were genetically diverse. The necessary diversity could be maintained only in reasonably large and widely distributed populations.

It appears that the territorial grassquits of Jamaica are very much less dense (by a factor of 2 to 3) than their social counterparts in Costa Rica. This is in accord with the supposition that the social form has a higher reproductive capacity. And since the grassquits are largely birds of old fields and forest edges, they are discontinuously distributed in Costa Rica (where fields are scattered) and continuously distributed in Jamaica (where fields are more often continuous). Consequently, the Costa Rican grassquits exist as small, relatively isolated colonies.

Very little is known about the degree of heterozygosity in natural populations of birds, which precludes predicting the degree of heterozygosity that might permit territorial behavior to evolve. However, we do know that both isolation and population size exert considerable influence on the degree of genetic diversity of natural populations. In very small populations, random drift can lead to fixation or loss of genetic loci. This decay of genetic variation is counterbalanced by the forces of mutation and immigration. There is evidence that for lizards (*Lacertilia* sp.) large population size and migration between adjacent populations is necessary for the maintenance of genetic diversity (Soulé, in press). Lizards from small, isolated island populations showed less variation in electrophoretically detectable isozymes than lizards from large island populations. The decrease in enzyme variation was correlated with a decrease in morphological variance. This result indicates that isolation and small population size result in a decrease in genetic diversity and could, therefore, limit the expression of such traits as territoriality.

A study of grassquits on an island smaller than Jamaica (Cayman Brac) revealed an intermediate degree of territoriality. The island's population was of a size suggesting an intermediate degree of heterozygosity. The fact that both social organization and genetic diversity do vary between islands of different size makes this explanation for the appearance (or nonappearance) of territorial behavior in different populations of the same species a viable one.

The second way in which we may apportion space is to allow mutual use of a large area by a group of individuals and to establish social priorities within the group to determine the use of a particular spot at a particular time. A band of howler monkeys (*Alouatta* sp.) (Carpenter, 1934) may, for example, range over a considerable area, no one member of the band retaining possession of a particular spot for more than a relatively brief period. Yet, at a given moment, all portions of the inhabited area are not equally accessible to all individuals. A subordinate juvenile must be contented with the places not claimed, for the moment, by his superiors. Social status, which is determined by a combination of physical (hormonal) and psychological (age and experience) factors, is thus the determining variable in the immediate apportioning of space.

What Determines the Density?

As with most other biological problems, in considering the factors that limit the densities a population can attain, one must distinguish between proximate and ultimate causes. We recognize that the ultimate explanation for seasonal migration in the white-crowned sparrow lies in the enhanced reproductive success of the migrants. The immediate explanation, however, has to do with light-induced changes in the pituitary, which lead to an increase in the output of gonad-stimulating substances (Wolfson, 1960). As our concern at present is with the immediate or proximate regulators, we can exclude consideration of food and climate. Where these are not directly perceived as limiting by the animals concerned, they usually cannot function as the proximate regulators. The qualification, "usually," seeks to avoid denial of the fact that, especially in microorganisms, the depletion of food supplies may be directly halting a growth in population density. This process has even been suspected in the case of a few vertebrates, the spectacular periodic crash of the populations of northern fox (*Vulpes*), for instance, having been attributed to a decline in the numbers of their prey, and the increase in foxes to a return of the prey to a dense population. The emphasis here is on the fact that in most populations a continued increase in food will still not allow an immediate and concomitant increase in density. Only through time and the selection of a new set of controlling variables can the density-control mechanisms of these vertebrates be altered, natural selection having already established limits that tend to prevent potential food supplies from being overreached.

In addition to food supplies acting as immediate regulators through starvation, it is conceivable that the perception of a particular amount of food or of a potential food store may influence reproductive behavior and, hence, density. The increase in clutch size of avian predators and of some gallinaceous birds in "good" insect or rodent years, an increase that antedates the actual availability of the food (Lack, 1954), suggests this possibility. Fortunately, since this is a problem not unamenable to an experimental approach, an eventual answer may reasonably be hoped for. In the meantime, we shall turn to the large numbers of cases in which some factor other than food or climate serves as the proximate regulator.

An animal may be territorial or it may not. If it is territorial, that is, is intolerant of the presence of its congeners (other than a sexual partner or its offspring), the maximum density of its population will be a function of the compressibility, or minimum size, of the territory. This is seen to be the case in a large proportion of bird species. Alternatively, if an animal does not show territorialism of this sort, it may still possess an "individual distance" within which the presence of congeners, if allowed at all, will occasion psychosomatic symptoms that greatly increase mortality. Thus the work of Christian (1960) and Chitty (1959), among others, has shown how crowding mice (various genera and species) may lead to adrenal hypertrophy and a concomitant increase in mortality. The mortality may result from an increase in aggressiveness, as in red deer (*Cervus elaphus*) confined in too small a space (Darling, 1937). It may stem from interference with reproduction, as shown by Helmreich (1960), whose mice produced a decreasing number of young as the amount of crowding to which they were subjected increased. It is also possible that the adrenal hypertrophy, associated with crowding, is a sign of the "stress syndrome," so flamboyantly elaborated by Selye (1956).

Another aspect of this type of physiological control has been suggested (Welch and Klopfer, 1961). In addition to mean adrenal weights (or the adrenal–body weight ratio) increasing with rising density, there is also a significant rise in variance (at least in adult, male, white Swiss mice). As a result, the probability of death from effects associated with adrenal hypertrophy is not shared equally by all members of the population.

> It appears that there are two predominant kinds of psychosocially influenced endocrine responses to crowding which oppose each other in the effect that they produce on the population. First, the true elevation in adrenal weight throughout the entire population together with the greater physiological homogeneity of the majority, favors the likelihood that a large portion of the population will experience the consequences of acute hyperfunction simultaneously. Second, this tendency is opposed by the increasing number of deviant individuals whose death will precede the others, and, reducing density, reduce the likelihood of a major population crash. [P. 259, Welch and Klopfer, 1961.]

If this situation proves generally true, it provides a method for a *selective* reduction in density and thus for the avoidance of a wholesale population crash. Welch and Klopfer (1961) have also suggested, although not proved, that the mice with the largest adrenal weights are those who are most closely related in the dominance hierarchy. Therefore a tendency toward ever more discrete hierarchical levels should be manifest in populations subject to periodic increases in density. (Siegel and Siegel, 1961, suggest that among domestic fowl the largest adrenals are to be found in individuals at both ends of the dominance hierarchy.)

A caveat does need to be entered here, however, for those who, extrapolating to man, see in this density-dependent response an automatic solution to human population problems. In a study designed to discover the nature of the stimuli that signal crowding Sattler (1970) isolated pairs of mice and provided some of them with odors derived from an overcrowded colony. These pseudocrowded mice showed the same degree of adrenal hypertrophy as those in the crowded colony but, unlike the latter, reproduced successfully. Their reproduction was identical to that of the central pairs (which received no colony odors), though the size of their adrenals differed. Hence the reproductive failure seen in crowded colonies may not be causally linked to adrenal malfunction.

Finally, if neither territorialism nor individual distance plays a role in the social organization, a proximate limit on density can be provided by the fact that suitable sites for the animals' habitation may not be infinite in number. Kluijver and Tinbergen (1953), for instance, found that an increase in the number of tree holes in a wood (specifically, of nesting boxes) allowed an increase in the number of nesting tits (*Parus major*). Presumably the shortage of suitable nesting sites was limiting the density below the level at which a purely adrenal consideration would have limited it. In birds such as gannets (*Sula* sp.) or puffins (*Fratercula* sp.), which require ledges or steep cliffs on which to breed, the actual amount of space available within a given colony may also impose a ceiling below the possible maximum. It is to be expected that the fewer the social responses in an animal's behavioral repertoire, the more likely that a physical factor, such as available space, will be the immediately limiting factor. In social species, behaviorally or physiologically imposed limits are first to assert themselves. One would then predict that maximum densities would be attained by invertebrates. To make a fair comparison between the densities of colonies of vertebrates and invertebrates, it is necessary to express the number of individuals in terms of some relative measure of area. Though this sort of comparison has not been attempted on any large scale, intuitively, the prediction seems likely to be sustained.

Far less attention has generally been devoted to the converse of the major question asked above: Are there minimum densities below which survival is not possible? The forms that can exist in a largely solitary manner, widely separated from their congeners, are relatively few in number. Aggregations above a certain density may be advantageous for a variety of reasons. Their value may lie in the enhanced protection afforded against either predators or environmental toxins. In the case of small numbers of fish, to whose water soluble toxins have been added, most may succumb, where as with a large number most may survive (Allee, 1931). Presumably this survival is due to the secretion of a neutralizing agent. That secretory products have social effects is known from a study relating the rate of the growth of tadpoles to the population density (Richard, 1958). A medium previously inhabited by dense aggregations retains this growth-inhibiting effect. However, the value of density may be in stimulating reproduction, denser colonies of certain birds, for example, tending to breed earlier than less dense colonies (Coulson and White, 1959). Frequently known as the Darling effect, this acceleration of the breeding period has been considered to be due to the additive effects of mutual stimulation induced by sexual display. Darling (1938) believed that the adaptive value of a shortened breeding period in colonially nesting birds was that it would provide for the selection of the aggregation tendency. Others, particularly Fisher (1954), believed that Darling's supporting data could more simply be interpreted to mean that younger birds, which usually make up the smaller and less dense colonies, have a later and more protracted breeding period because of their age. Both Fisher and Darling may have been correct. In the kittiwake (*Rissa tridactyla*), at least, the older birds do breed earlier than younger birds, but the greater the density of the colony, ages being equal, the more the breeding period is accelerated. There is, however, no contraction of the total breeding period as Darling believed. Whether a selective advantage accrues to reproducing early in the season may thus depend on climatic conditions. Nonetheless, the essential mutual enhancement of reproductive tendencies that one can observe in social groups assures the maintenance of a density that varies between both a maximum and minimum value.

Changes in density as the result of the imposition of artificial conditions and the relaxation of natural selection may have interesting side effects. It is to be expected that animals domesticated for long periods, whose food supply is independent of their density, will no longer show either the endocrinological or psychological effects of crowding to which their wild counterparts are subject. Man's selection of the hardiest stock over many generations will have assured this insensitivity of domestic animals to crowding. But what about human populations that, generally, have not been subjected to the rational types of selection that our domestic stock characterize? What can be expect of an organism that originally evolved under conditions of relatively low density (cf. Bartholomew and Birdsell, 1953) only to rapidly move into conditions of extraordinarily high density? It may be that a selection against individuals unable to bear the stress of high population densities has been occurring. Certainly, other similar kinds of selection are known to be operative, for example, the selection against the Y-bearing sperm in the case of airmen traveling faster than sound (during the two months prior to conception; Snyder,

1961) or the increased mortality of those teenagers whose personalities include a need to drive wildly (John Calhoun, personal communication). Given man's procreative abilities and medical skill, however, it seems equally likely that individuals unable to tolerate high-density stress will nonetheless reproduce their kind. Indeed, the neuroses such stresses can produce may even lead to a compensatory hypertrophy of sexual and reproductive activity. Thus a gradual change in the temperament or behavior of the human species might occur, a change that is basically maladaptive and that will become increasingly so as human populations continue to grow.

It is possible, of course, that ontogenetic factors may determine the degree to which crowding can be tolerated, both psychologically and physiologically. The city-bred boy who is frightened by the "emptiness" of the country is well known to us, although it is not certain whether the transplanted city boy is not in fact in a healthier situation, his subjective attitude notwithstanding. However, in at least one organism, an early experience of crowding can make subsequent crowding an unobjectionable condition. Ellis (1959) found that young locusts of two species, *Locusta migratoria* and *Schistocerca gregaria,* were inclined to aggregate or not depending on whether they were reared singly or in groups. The group-reared animals aggregated readily; those reared singly required several hours of habituation to their confreres before they would group. Curiously, the definition of group rearing, insofar as the hoppers' perceptual apparatus is concerned, is in terms of tactile stimulation. Hoppers reared in transparent boxes, in full view, smell, and hearing of groups of hoppers, showed only a slight increase in aggregation tendency in *Locusta* and none in *Schistocerca.* Stimulating isolated hoppers, by touching them repeatedly with fine wires, however, produced aggregation tendencies of nearly normal proportions. (No implications for *Homo sapiens* need be drawn from this information.)

In any event it appears possible to determine whether ontogenetic experiences can modify the effects of crowding. The considerable practical value to such studies should also be apparent.

THREE

Why Does Species Diversity Vary?

A traveler progressing from either pole toward the tropical regions about the equator cannot help but note a general increase in the numbers of different kinds of plants and animals. Why should this be so? Why should the number of different species of organisms not be uniform throughout the globe? Or why should there not be a random variation? Why is it that there is a cline with a maximum concentration of forms in the equatorial tropics?

Needless to say, there are other patterns of abundance that could also be observed by a perceptive traveler. In isolated areas or small islands far from the shores of continental landmasses, the number of species will be reduced (see MacArthur and Wilson, 1967). Other systematic variations in species abundances can be noted as well. Superimposed on all other patterns, however, is that which is probably the most conspicuous pattern of all, namely, the increase of species diversity toward the equator (MacArthur, 1965; Pianka, 1966; Fischer, 1960).

Let us examine this problem more closely. Upon doing so, we shall observe, first, that the increase in the number of species that occurs toward the equatorial regions is not an increase that is uniform for all families or classes of animals. In the case of mammals, for example, there is an undoubted increase in the number of flying forms but no increase and possibly even a decrease in the number of subterranean or burrowing forms. In the case of birds, there is a difference in the way in which the number of species changes for the passerines, or perching birds, and the nonpasserines. We shall soon consider the specific details of this difference. Second in at least some instances the number of species of a particular group increases rather abruptly at a particular latitude rather than regularly, as would be the case with a true cline. This fact must be borne in mind when an explanation is given for the enhanced faunal diversity of the tropics. Finally, there are a few groups, freshwater diatoms among them, that appear not to increase their diversity in the tropics (Patrick, 1964). Exceptions require explanations no less than do rules.

In attempting to interpret and analyze the causes of the enhanced diversity of tropical regions it will be convenient to group explanations into three categories: those based on considerations of time, ecology, and, most important for our present purposes, behavior.

TEMPORAL FACTORS IN DIVERSIFICATION

Let us imagine a large aquarium into one corner of which we introduce a few crystals of potassium permanganate or some other soluble dye. After this material dissolves it will diffuse throughout the aquarium, and, if we continually add new crystals to the original pile, we shall find that however much of the material diffuses outward from the center there will still be a region of greatest density that grades imperceptibly to the regions of lowest density. If we were to look upon the tropical regions as those regions within which species have been and are being created to a much greater extent than other places on the globe, we would have a ready explanation for the fact that the tropics are indeed richer in species than other regions. This is a point stressed by Darlington (1959). Periodic climatic catastrophes that have decimated living organisms have been far more common in the higher latitudes than toward the equator, and thus there has been, in effect, much more time for the processes of species creation to occur in the tropics than elsewhere. In addition, a variety of other reasons have been proposed that would endow the tropics with a capacity for a greater rate of species production (Darlington, 1959). These reasons, such as enhanced mutation rate, shorter generation times, and more intense selection pressures, can be regarded as being based primarily on a temporal factor. It is clear that in an infinite, or at least a very long, period of time some sort of equilibrium is going to be attained that will reduce the steepness of the species gradient considerably. Similarly, in our aquarium the solution will eventually become saturated, and there will then be a minimum difference between the amount of the permanganate at the center of diffusion and at the extreme periphery. Returning to the natural situation, if the periodic climatic catastrophes continue to occur more frequently in polar regions than in tropical regions so that, in effect, species are eliminated as they move out from the tropics, the cline could be maintained. Then evidence of a high rate of extinction in higher latitudes should be sought.

However, there are several objections to these explanations of enhanced tropical diversity. To begin with, there is some evidence to indicate that there is not a cline between tropical and polar regions but rather an abrupt discontinuity. The number of breeding birds in homogeneous woodlands does not differ by any order of magnitude between Nova Scotia, New York, and Florida. There is roughly the same number of bird species in each of these three areas. Suddenly, however, to the south of Florida, one crosses an invisible line, and the number of breeding bird species increases enormously; from 100, 200, or 300, one suddenly goes to 700, 800, or 900. It is unfortunately not possible to say at this time whether the same thing is true when one proceeds from South America northward to the tropics. Di Castri (1967) found it so for some soil arthropods. However, data from Patagonia collected by Vuilleumier (in press) suggest that bird species diversity in temperate South America may depend on different factors than in North America. Elsewhere, it does still look as if the region of an enhanced faunal diversity, at least with respect to birds, is a region of a rather high climatic stability that does not grade into adjacent regions.

Figure 3–1, by way of illustration, depicts the landmasses of the New World that can be characterized as climatically stable. By stable we mean that there is no frost, that the rainfall varies very little from

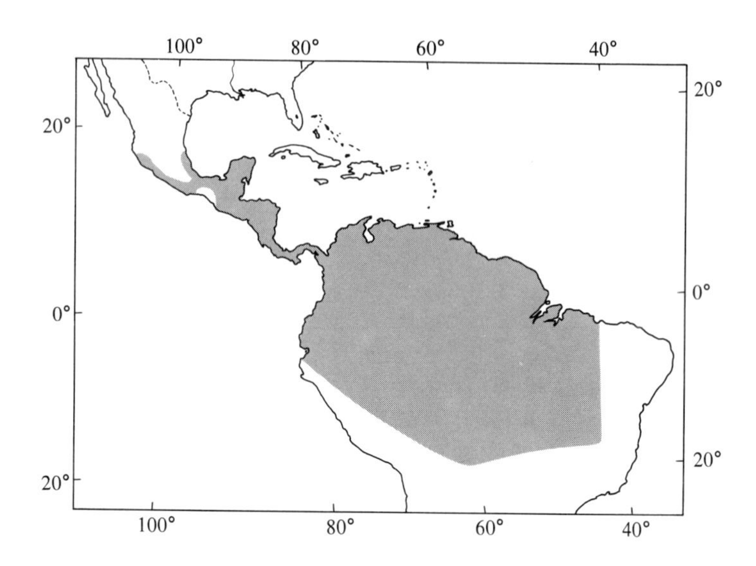

Fig. 3–1 Regions of maximum climatic stability. The portions of the shaded areas that are at sea level have the following characteristics: (*a*) no frost, (*b*) monthly rainfall variability less than 20 percent of annual variability, (*c*) average daily temperature range greater than average annual range. (After Klopfer and MacArthur, 1961.)

month to month, and that the annual variation of temperature is no greater than the daily range. Within the area encompassed by Fig. 3–1 there is a very large number of breeding birds. To the north the number drops appreciably, and across the two points censused it drops suddenly, disjunctly, not continuously. This is clearly not the situation that one would expect to obtain if the number of bird species was purely or even primarily a result of a diffusion process dependent only or largely on time. This observation, if it can be confirmed for other regions shown on this map, would tend very much to weaken the diffusion explanation.

The second difficulty lies in the fact that the diffusion or temporal explanation provides no clues as to why there should be differences in abundances among different taxa. Why, for example, are individual passerines relatively more abundant to the north and nonpasserines in the tropics? Why are there so many more arboreal and volant forms of mammals in the tropics and so many fewer subterranean and purely terrestrial forms? These facts call for some other explanations. Whatever role time has played, and clearly time has played some role in determining differences in faunal abundances, it is unlikely to have been the sole determinant of faunal distribution and abundances.

Temporal factors can be important over shorter stretches of time, too. If competition between sympatric species is of consequence only during particular portions of their life cycle, then a shift in the phase of one or the other cycle of two potentially competing species can mitigate the effects of their conflict. In the case of birds, at least, it is during the breeding period that competition (for both food and nest sites) reaches its peak. Thus it is of particular interest to note that in the tropics many related species have staggered their breeding cycles. The relevant data, culled from Skutch (1954, 1960, 1966, 1967) are summarized in Fig. 3-2.

Incomplete as the data are, it will be noted that the breeding seasons of these species extend over many more months than is possible to the north and that the peaks for the various species come at different times. This process, in effect, extends the temporal dimensions of the total "volume" that has to be divided among all the contending species. It allows for an increased number of species without requiring a corresponding decrease in the size of individual niches.* It does require, however, that the onset of breeding activities be rather precisely controlled for each of the potentially competing species.

In addition to the climate of the tropics allowing a staggering of breeding activities, it also allows for a lengthening of each period. This might be thought to have a bearing on total productivity and hence on diversity (see Connell's argument below), but Ricklefs (1966) has disputed the importance of a temporal component of diversity in tropical birds. His work indicates that all species increase their breeding times in a proportional manner, thus negating whatever effect the enlarged opportunity might provide. His data, far more complete than those presented in Fig. 3-2, also directly challenge that result, providing no evidence that related sympatric species stagger their nesting season. Ricklefs does suggest that the longer breeding periods of the tropics allow for a reduced intensity of breeding activities, which, together with the smaller clutches of the tropics, could reduce interspecific competition. Apparently, this matter can be resolved only by additional data. Nor will the outcome of further studies necessarily be to support either of the above notions; Miller (1959)

* For the definition of *niche*, see p. 75.

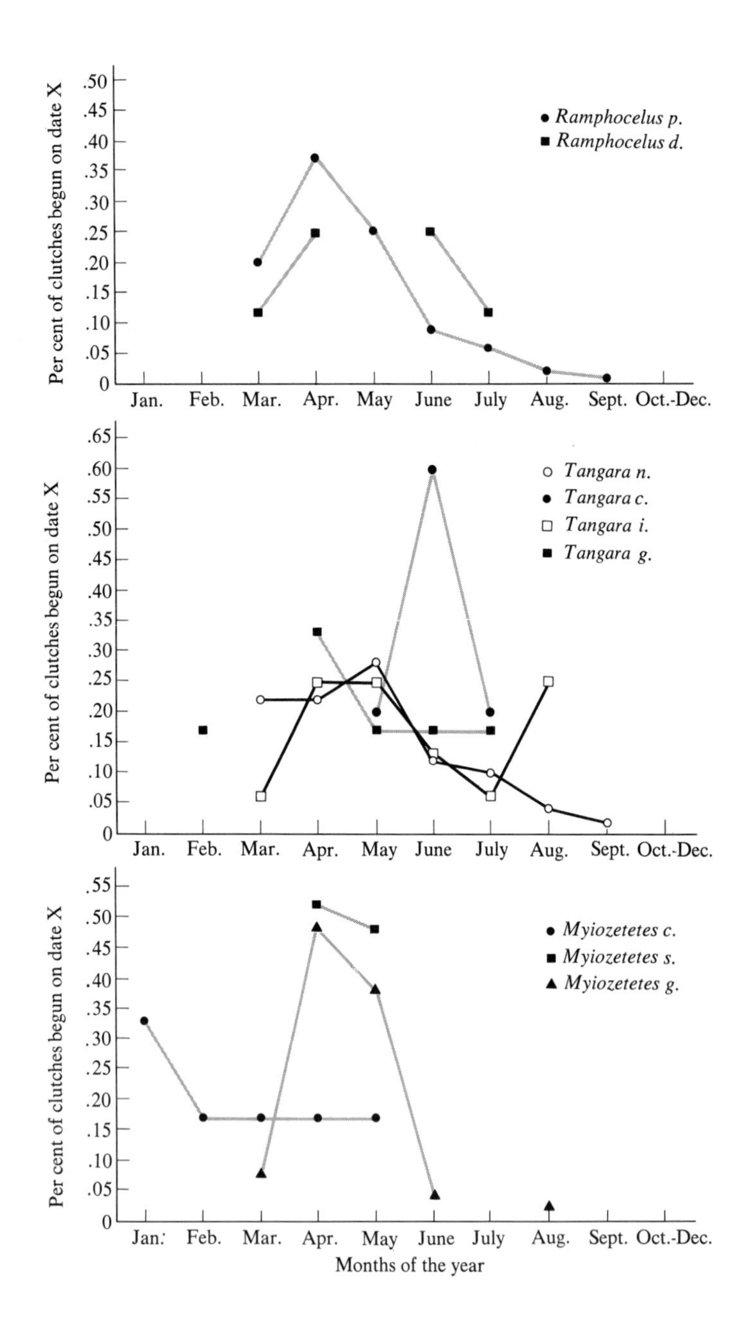

Fig. 3–2 Breeding seasons for three groups of related and sympatric species of tropical passerines. Note the seasonal differences in the time of the peak of the breeding season. In temperate-zone species there is rarely such a spread.

has evidence suggesting a six-month breeding cycle in at least one equatorial sparrow.

Finally, it has been argued that the climatic stability of the tropics could lead to a loss of reproductive synchrony between non-contiguous populations of the same species. The ensuing reproductive isolation could then promote the creation of new species (Fedorov, 1966). There has been a strongly stated suggestion that many conventional species in fact consist of reproductively isolated aggregates, which are not therefore acknowledged to be distinct species (Ehrlich and Raven, 1969). Hence the problem involves establishing empirically the degree to which tropical species are asynchronous (if at all) and then resolving the issue of whether reproductive isolation is a necessary and sufficient condition for speciation (also see Mayr, 1963).

ECOLOGICAL FACTORS IN DIVERSIFICATION

Under the second category of explanations, ecological causes, fall such factors as topographic variability, habitat complexity, population interactions, and total niche volume available. Miller (1942) has argued that the number of species or subspecies into which a genus or species could be divided was very much a function of the topographic variability of the range that the taxon inhabited. Where there was great uniformity of range, there was a relatively small tendency for the taxon to break up into smaller semiisolated units. Where topographic diversity was somewhat greater, provided the species was endowed with whatever capacities are required to enable natural selection to produce local differences, a certain degree of reproductive isolation would follow. As the result of genetic drift and adaptation to local conditions a separate subspecies or species might be formed. Now, if it could be shown that tropical regions were, first, more extensive geographically than other regions or, alternatively, were topographically more diverse than other regions, we might here have at least a partial explanation for the distribution that we have noted. There does not, however, seem any a priori reason to believe this to be the case. Despite the evident validity of Miller's argument, there is nothing about either the areal extent of the tropics in the New or the Old World or their topographic diversity that would account for the great increase in faunal diversity.

Another possibility that we might consider is based on the clear fact that the floral substrate of the tropics, if not topography, is in fact more complex than that to the north or south. Orians (1970) points out that tropical forests offer a greater range of resources. In part, of course, this consideration begs the question for it does not give us any clues as to why the floral diversity is so rich in the tropics. However, although we tend to believe that whatever explanations apply to faunal diversity will apply to all or at least most groups of animals, we are not constrained to believe that the same explanation must also apply to plants. Their taxonomy and competitive relations differ rather significantly from those obtaining among animals. The botanical concept of species, because of the high incidence of polyploidy in the plant world, must differ somewhat from the zoological species concept. Similarly, the comparative lack of motility and the largely autotrophic habits of plants eliminate most forms of prey–predator relations. This does not deny the role of animal "predators" on plants, which, as Janzen (1970) has shown, does influence tree diversity (also note Harper, 1969). Thus if we accept as a fact the enhanced complexity of the floral substrate of the tropical regions, complexity both in terms of species composition and complexity of layers, it is possible to argue that this difference from the northern and southern temperate flora provides for many new microhabitats or niches for animals. Therefore it should allow for many more animal species.

To determine more precisely the importance of the complexity of a habitat, it has been useful to distinguish between the total faunal diversity of an area and the diversity of the more or less uniform habitats that together comprise the area. Thus the total area of Jamaica has a particular degree of diversity (labeled gamma diversity by Whittaker, 1960, or between-habitat diversity by MacArthur, 1965), while the grasslands, old fields, secondary forest, cloud forest, and so forth have a distinct measure of diversity (the within-habitat diversity).

To specify the degree of diversity quantitatively, it has proved useful to employ the measure of uncertainty in information theory. This can serve to both specify the diversity of birds in a woods and the diversity in plant species by the distribution of foliage densities. Foliage density diversity (variations in density with height), in fact, appears the best predictor of bird species diversity (MacArthur and

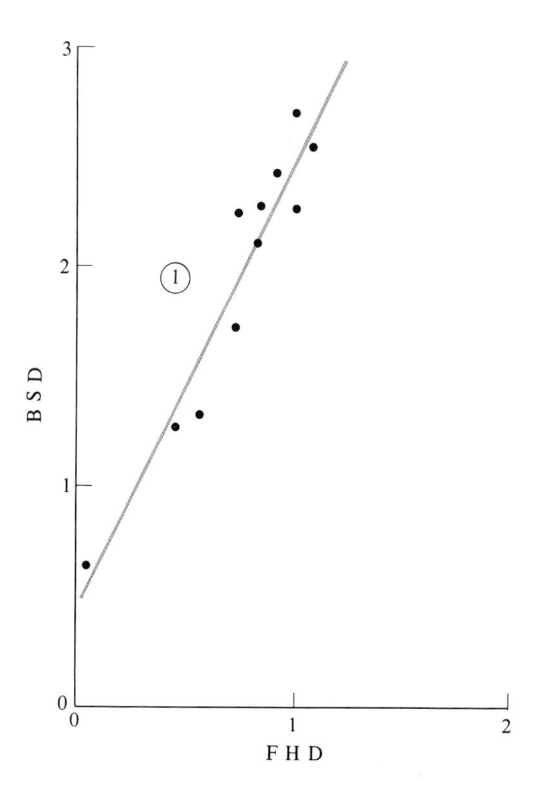

Fig. 3–3 Bird species diversity (BSD) as a function of foliage height diversity (FHD). (1) represents a single tropical habitat. (After MacArthur and MacArthur, 1961.)

MacArthur, 1961; Klopfer, 1960a). Figure 3–3 shows the relation between foliage height diversity and bird species diversity found by MacArthur in a variety of habitats.

It must be noticed that Fig. 3–3 is based on analyses of habitats in temperate North America, although Recher (1969) found a similar relationship in Australia. If one makes similar analyses of tropical habitats and plots these points on the same graph, one discovers that the line that is formed is not simply an upward extension of the original plot but rather is displaced above the original plot (MacArthur and MacArthur, 1961). What this graph means is that even in areas of equivalent floral complexity and diversity the tropical fauna is still richer than that to the north. Specifically, the number of breeding birds in a tropical rain forest is more than that predicted in terms of the increase in floral complexity of the tropical rain forest over a temperate forest. In a tropical savannah, where floral complexity is no greater than that obtaining in a temperate zone savannah, there are still many more birds. Now, we do not mean to deny the obvious fact that the regions of the tropics are indeed much more complex floristically, in diversity and stratification, than plant communities elsewhere. Thus the number of microhabitats available in the tropics is indeed greater than the number available in areas of equal size elsewhere. Bromeliads, for instance, provide a microhabitat for a certain species of mosquitoes (family Culicidae) that is totally lacking in other kinds of forests. The tightly closed canopy of a rain forest certainly has no close equivalent in the north, but the number of bird species, and presumably the number of species of other kinds of animals as well, increases out of all proportion to this increase in floral complexity.

One response to this assertion has been that the increased tropical diversity is indeed attributable to the between-habitat differences (gamma diversity) and thus to habitat diversity. For instance, MacArthur (1965) and his students found that the within-habitat diversity of the island of Puerto Rico, which has relatively few species altogether, was almost as great as that of the tropical mainland with its much greater total diversity. MacArthur concludes that the greater total diversity of the tropics is thus due to "a finer subdivision of habitats (habitat selection) more than by a marked increase in diversity within habitats." In the absence of an operationally useful definition of habitat, this explanation begs the question. The fact is that fields and forests of the tropics, whether we consider them as single or multiple "habitats," do generally contain a greater number of species than fields and forests to the north. Are the behavioral attributes of the inhabitants of relevance to an explanation, or are strictly vegetational—physiographic factors a sufficient cause?

One rather different ecological argument to explain tropical diversity should be considered. It is based on the assumption that with greater climatic stability less energy is needed by individual organisms for physiological regulation (for instance, temperature control) and more is available for reproduction. This implies larger populations, greater genetic variety, and thus more speciation (Connell and Orias, 1964). There is more to the argument, of course, especially with respect to evidence supporting the primary assumption. However, it is clearly inapplicable to birds. Data from Skutch (1966, 1969) show that tropical birds have a much reduced breeding success compared with those from temperate zones, in addition to substantially smaller clutches and a longer developmental period. This is in accord with the impressions of many naturalists.

It is probably true, however, that the greater photosynthetic productivity of the tropics and reduced seasonality allow for marginal levels of subsistence that might not otherwise exist (MacArthur, 1969).

BEHAVIORAL FACTORS IN DIVERSIFICATION

If within a fixed area we wish to increase the number of species, it is hypothetically possible to do so either by increasing the amount of time during which speciation can occur or by increasing the "space" within which niches can be carved out for different species. This niche "space" must be thought of as a hypervolume whose parameters are all those variables whose limits determine the presence or absence of a species (Hutchinson, 1957). The number of species can also be increased by reducing the size of the niche that the individual species requires. When we talk about a smaller niche, what we mean is that the measurable range of conditions that determines the presence of *a* is less than the range that determines the presence of *b*. Concerning the argument that the greater number of species in the tropics is to be accounted for in terms of smaller niches possessed by tropical species, we shall focus first on some of the behavioral aspects of a reduction in niche size.

One line of evidence, suggesting that it is a decrease in the volume of individual niches that accounts for the enhanced faunal diversity of the tropics, comes from a study by Klopfer and MacArthur (1961). Their data indicate that a reduction occurs in the amount of character displacement among the sympatric species in tropical regions. Wherever two species are found living together, feeding upon similar kinds of food or utilizing similar kinds of breeding habitat, there is a difference in their size or in the size of certain of their appendages, which tends to make the individuals of the two species pick food objects or utilize nesting sites of different sizes. This reduces or eliminates competition between the species for those particular elements of the environment, which usually are those that limit the expansion of the species' numbers. As has been suggested elsewhere, this type of diversification of the demands made on the habitat by species inhabiting similar regions is a fairly widespread phenomenon, particularly among vertebrates. Hutchinson (1959) has indicated that in these cases of sympatry, where two species inhabit the same area simultaneously, the ratio between the size of the larger and the smaller will be 1.2 to 1.4. That is, this is the ratio obtained when one compares trophic characters such as the length of the bill or the mandibles. Hutchinson has compared a variety of forms, invertebrate and vertebrate, and in all cases found that the character that determines the nature or size of the food is different in size to the extent indicated by this ratio.

It is a curious fact, however, that when one compares the amount of character displacement that has occurred among sympatric forms in the tropics a much reduced ratio is obtained. In Table 3–1 are some data from Panama collected by Klopfer and MacArthur (1961) that indicate the differences in the sizes of the bill of a group of birds commonly found feeding together. The character displacement ratio in a good many instances, it will be noted, is unity, suggesting that in these species, at least, the food portion of the species niche overlaps considerably if not completely. If this is the case, unless the entire premise from which the arguments for niche diversification stem is invalidated, it follows that the exclusive portion of the niche of each of these species must be much reduced in volume.

A far more detailed analysis of differences in bill size has been completed by Schoener (1965). His findings implicate a variety of other factors in the determination of differences. Large ratios in size were seen among families whose members feed on food of low abundance and those found in sympatric congeneric associations on islands, especially small islands. Schoener's data largely support the idea that bill-size differences play an important role in partitioning food but offer no particular support for the Klopfer–MacArthur (1961) view on the significance of differences in ratios between tropical- and temperate-zone species. On the basis of this evidence, it remains an open question as to whether the niches of tropical species are smaller.

What do we imply in terms of behavior when we say that the volume of the niche of a species is reduced? Essentially, we mean that the behavior of the animal has become more stereotyped. When the niche is reduced the range of objects in the environment to which the animals responds by feeding or nesting or taking shelter is reduced. Its behavioral repertoire may be reduced and so may the range of different kinds of responses that can be elicited. Therefore one way of determining whether or not tropical niches are indeed smaller is to determine whether or not the behavior of tropical species is, by and large, more stereotyped and less plastic than the behavior of species to the north and the south. Let us examine some of the evidence that exists concerning this question and then con-

TABLE 3–1

Character Displacement Among Sympatric Species of Birds of Panama and Costa Rica

	BILL LENGTHS, ADULT MALE	
	MEAN VALUES	RATIO OF BILLS OF LARGE TO SMALL SPECIES
Ramphocelus passerinii	13.5	
		1.01
Ramphocelus dimidiatus	13.7	
		1.11
Ramphocelus icteronotus	15.2	
Throupis episcopus	12.4	
		1.06
Thraupis palmarum	13.2	
Tangara inornata	8.9	
		1.09
Tangara nigro-cincta	9.7	
		1.02
Tangara icterocephala	9.9	
		1.04
Tangara chrysophrys	10.3	
		1.04
Tangara gyrola	10.7	
Myiozetetes cayenensis	13.9	
		1.01
Myiozetetes similis	14.0	
		1.00
Myiozetetes granadensis	14.0	

sider other kinds of evidence that have not yet been collected but that would throw light on it, indicating whether our proposed solution is valid or not.

Let us turn first to the way in which the relative numbers of passerines and nonpasserines change in the tropics. Table 3–2 indicates how the proportions of nonpasserines change from north to south. In general, these figures support the conclusion that toward the equator conditions are favorable for the existence of nonpasserines to a greater degree than farther north. Tropical and temperate areas certainly have had rather different geographic and

TABLE 3–2

Effects of Latitude on Passerine–Nonpasserine Ratios*

REGION	PERCENT OF INDIVIDUALS THAT ARE NONPASSERINE†	PASSERINE–NONPASSERINE	TERRITORIAL MALES PER 100 ACRES PER SPECIES: NONPASSERINE	TERRITORIAL MALES PER 100 ACRES PER SPECIES: PASSERINE
Ontario	6.2	5.76	1.7	9.8
Maine	1.2	4	3	12
Maine	3	3.97	3.1	12.3
NW Territory	8.8	3.18	5	15.9
NW Territory	4.5	4.59	3.4	15.6
Idaho	12.4	1.68	10.8	18.1
South Dakota	20	1	15	15
Colorado	8.8	4.79	3.3	15.8
Colorado	9	5.29	0.7	3.7
Oregon	5	2.38	5	11.9
North Carolina	0.4	23.77	1.3	30.9
New York	8	3.83	1.8	6.9
Ohio	5.5	6.46	1.3	8.4
Maryland	9.8	5.33	8.6	45.8
Tennessee	5.4	4.38	5.8	25.4
West Virginia	2	4.07	6.7	27.3
North Carolina	11.4	2.68	2.5	6.7
North Carolina	3.3	3.88	2.5	9.7
Georgia	12.5	2	1	2
California	13.3	2.50	9.6	24
North Carolina	3.5	2.18	12.5	27.3

* Derived from *Audubon Fieldnotes*.
† Exclusive of ducks, grebes, and other aquatic species.

climatic histories. However, as indicated earlier, historical factors could explain the changes in the proportions of nonpasserines between Mexico and the northern United States and Canada only if the nonpasserines have been restricted to the tropics or the passerines to the north while speciation was taking place. Perhaps this was the case, but the fact that nonpasserine land birds are restricted to the tropics throughout the world, and not just in the New World, suggests that it is more than a mere physical barrier that prevents them from colonizing the north temperate areas. Instead it is rather reasonable to assume something else, namely, that nonpasserines

TABLE 3–2 (continued)

Effects of Latitude on Passerine–Nonpasserine Ratios

REGION	PERCENT OF INDIVIDUALS THAT ARE NONPASSERINE	TERRITORIAL MALES PER 100 ACRES PER SPECIES		
		PASSERINE–NONPASSERINE	NONPASSERINE	PASSERINE
South Carolina	9	3.66	3.5	12.8
Illinois	18	3.66	3.5	12.8
Georgia	15	2.07	5.6	11.6
Florida	16.7	2.00	8	16
Arkansas	10.7	1.74	3.9	6.8
Arkansas	7.1	3.28	5.8	19
Texas	20	1.76	6.3	11.1
Oklahoma	0	—	—	7.7
Wyoming	6	5.68	2.5	14.2
Texas	14	1.85	2	3.7
California	6	4.68	2.2	10.3
California	35	0.52	14.3	7.4
Arizona	37.5	1.27	2.6	3.3
Durango, Mex.	35	2	0.7	1.4
Zacatecas, Mex.	21.4	2.20	4	8.8
Chiapas, Mex.	43.5	1.60	6.5	10.4
Veracruz, Mex.	48.5	1.21	4.2	5.1
Veracruz, Mex.	31.4	1.01	7.4	7.5
Quintana Roo, Mex.	24	1.12	6.6	7.4
Veracruz, Mex.	34.2	1.47	5.8	8.5
Chiapas, Mex.	33.3	1.66	5.9	9.8
Veracruz, Mex.	35.3	1.48	5.2	7.7
Chiapas, Mex.	25	3.45	3.9	10

(and in our consideration we are limiting ourselves to terrestrial species, which obviously pose special problems) are of more ancient vintage than passerines.

If we make this assumption, one that is borne out by paleontological evidence, we may than continue our argument by assuming that the phylogenetically younger passerines have a less limited central nervous capacity than the nonpasserines and are thus more capable of modifying their behavior to fit changing environmental stimuli. This assumption seems to be supported by the fact that within a great many different lineages the tendency in evolution has been for flexible or plastic behavior to be promoted over highly stereotyped and inflexible response patterns. Now, if these assumptions are indeed valid, then we should expect the passerines to be relatively more abundant in unstable environments where their plasticity allows them to make the appropriate responses, responses that are ever changing with the seasons. These responses would be less readily possible among the more stereotyped nonpasserines. Furthermore, with less stereotyped behavior, the passerines should require larger niches. Thus if an analysis of census data were to show that the proportion of nonpasserine individuals does not increase toward the tropics, we would have to consider our notion of smaller niches in the tropics unsupported, or, alternatively, we would have to reconsider our initial assumptions. But the relationship between passerine and nonpasserine individuals as one approaches the tropics does, as we have seen, show the predicted change—the nonpasserines do become more common—and thus some support for our notion can be claimed.

With a reduction in niche size, that is, increased specialization or behavioral stereotypy, one would also predict that there would be a reduction in the number of individuals per species for a given area. This prediction follows from the following argument: If the requirements of an individual are general rather than limited and specific, that is, if the niche is relatively large, then many more individuals of that particular species can coexist in a particular geographic area. By way of example, consider a bird whose feeding is restricted to a single food plant. The food dimension of its niche is quite small. Far fewer individuals of this species can coexist in a limited area than would be the case of birds feeding from a larger variety of plants. Table 3–2 indicates how the number of individuals per species varies between passerines and nonpasserines in the tropics and nontropics. It is seen from this table that the mean abundance per species per 100 acres is reduced in the tropics for the passerines much more than for the nonpasserines. In Fig. 3–4 we have plotted the ratio of the mean abundance of passerines to the mean abundance of nonpasserines, and this ratio also drops in tropical regions. The ratio is greater than 1 at all latitudes, indicating that the nonpasserines have a lower mean abundance per species wherever they occur. This evidence suggests even more strongly that the passerine species are more able to expand into temperate habitats than are their nonpasserine counterparts. In turn, this suggestion does imply more stereotypy on the part of the nonpasserines, reduced niche

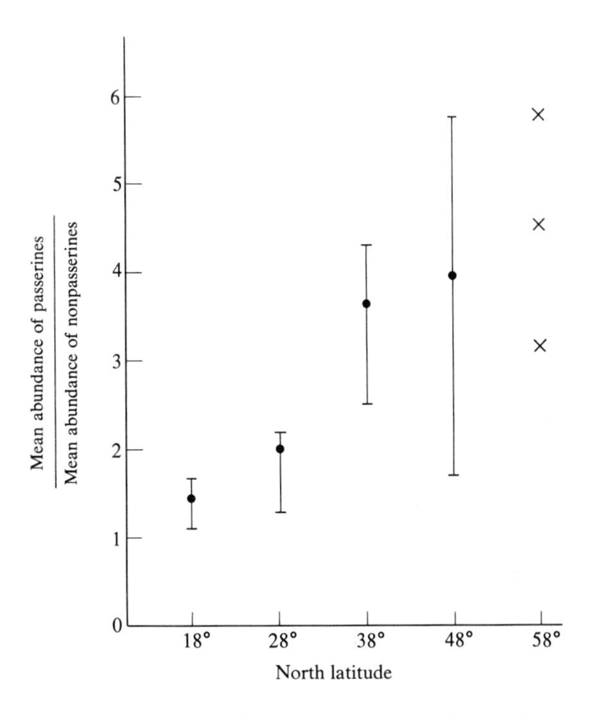

Fig. 3–4 Median and 95 percent confidence intervals for the median of the ratio, mean abundance of passerines–mean abundance of nonpasserines, for censuses grouped in 10-degree intervals from various latitudes. The ×s refer to values from single censuses. This graph shows that nonpasserines have lower mean abundances (that is, smaller niches) than passerines at all latitudes and that the passerines become less abundant toward the tropics than they are farther north. (After Klopfer and MacArthur, 1960.)

sizes, and, consequently, the possibility that many more species of nonpasserines can inhabit a particular area. Indeed, it has been pointed out (E. Simons, personal conversation) that a rather similar situation obtains in primates in which the specialized, primitive species are confined to the tropics, while the plastic species, more flexible in their behavior and more recently evolved, are ranged more widely. In summary, we can say that as one proceeds from north temperate regions toward the tropics the proportion of nonpasserines in the avifauna increases. The number of individuals per species for a given area decreases toward the tropics for the passerines but not for the nonpasserines, and at all latitudes the nonpasserines are less abundant than the passerines. We interpret this information as support for our notion that the phylogenetically older nonpasserine species are insufficiently plastic in their niche requirements to colonize temperate areas. In the tropics, niches are smaller and less subject to seasonal or other change.

Much the same point has been developed by Sanders (1969) with respect to marine invertebrates: Areas of stability have higher diversities. It is necessary, however, to distinguish between short-term and long-term climatic stability. The former refers to periods of stability lasting over no more than five generations in areas of normally unstable (i.e., unpredictable) conditions (Slobodkin and Sanders, 1969). Normally, the unpredictable, historically variable environments provide havens only for a few residents and others, perhaps more numerous, species from the fringes. These can be replaced (by immigration) when the inevitable shift in conditions leads to local extinction. Should there be several seasons of stability, however, the few more permanent residents may be expected to increase at the expense of the transients. The result is a paradoxical *decrease* in diversity with increasing (short-term) stability. However, long-term stability in marine environments does produce an increased diversity, in a fashion similar to that postulated for the birds of the tropics:

> Under benign conditions of long-term predictability physical conditions no longer are critical in controlling the success or failure of the species. This, in turn, allows the more subtle distinctions of the biological environment to be realized. Evolution is now in the direction of reducing biological stress (intense competition and predation, nonequilibrium conditions in prey–predator relationships, simple food webs, etc.).

> With time, such biological instructions give rise to biological accommodation. . . resulting in stable, complex buffered assemblages of many stereotopic species. [P. 91, Slobodkin and Sanders, 1969.]

But what about direct evidence for an increase in the degree to which tropical species are stereotopic or stereotyped? Evidence such as that of Terborgh and Weske (1970) that tropical species are vertically confined to narrower zones is supportive but cannot be conclusive. Speculation will, for a time, have to substitute for data.

If it is true that the enhanced faunal diversity of the tropics is, in the first place, due to a high degree of behavioral stereotypy and if it is further true that this behavioral stereotypy has certain neurological correlates or causes, then neuroanatomic investigations should confirm that tropical species differ from their temperate-zone counterparts with respect to their potentiality for plastic, nonstereotyped behavior. As Broadbent (1958) has pointed out, if a particular step in a behavioral sequence becomes the stimulus for a subsequent step sufficient to obviate the need for any kind of external stimulation, then the neural capacity required by the animal for the performance of the sequence can be reduced. This is merely another way of arguing that stereotyped behavior is more economic of neural tissue than behavior that is less stereotyped and requires a greater amount of information from the external environment. Nissen (1958), in fact, has pointed to much the same thing, so that the notion that stereotyped response patterns require a less elaborate central nervous system is not a particularly new or a particularly wild piece of speculation.

It is of interest, therefore, to compare the size and complexity of the brains of tropical and nontropical species. In this connection, the first thing to be pointed out is that, other things being equal, the best measure of relative neural complexity is the number of neurol elements contained within the central nervous system. The more elements, the more possible new connections. An interesting experiment by Fankhauser (1955) has shown that in a species of salamander the diploid form, whose cells (including brain cells) are smaller but more numerous than those of a triploid form, is capable of learning responses in a maze more rapidly. That is, with similarly designed brains, the animal with the larger number of brain cells is the more efficient at modifying its responses. This experiment certainly tends to support our contentions. Further evidence in this direction is afforded by the work of Rensch (1959), who has shown that the smaller of a pair of animals of the same species (but of a different strain) is generally less able to learn as wide a variety of tasks or to learn them as rapidly as his larger counterpart. Insofar as the number of brain cells is concerned, the absolute number is greater in the larger animal. Here again, the larger animal is endowed with a greater potential for behavioral plasticity because of its increased number of neural elements and, just possibly, their increased size. These facts become relevant when one takes cognizance of the well-known generalization that, by and large, tropical homothermic vertebrates tend to be smaller than their northern and southern counterparts. Such a decrease in size increases the ratio of surface area to mass. This relationship between latitude and size has generally been interpreted as an adaptation to minimize heat loss in areas where it is a problem and to accelerate heat loss in the warmer parts of the globe where heat accumulation can be troublesome. Now, this generalization does have innumerable exceptions, but to the extent that it does apply it assures that the northern counterparts of tropical species are endowed with a larger number of brain cells and thus a greater potential for behavioral flexibility.

Margaret Mead (personal communication) has pointed out that an organism's susceptibility to an experimentally induced neurosis is in large part a function of the degree to which this organism has been accustomed to radical changes in its environment. She points out, for example, that, in the past, English children at age two years were suddenly given cold milk rather than the warm milk to which they had previously been accustomed. These children, however, had become thoroughly used to sudden changes according to fixed schedules, and this particular change therefore elicited no disturbance. For children reared under more permissive schedules, where rigid timetables are not adhered to, such a sudden transition can provoke a severe trauma. The analogy of this situation to animal studies, where a lack of predictability or change in the degree of predictability with respect to the stimulus situation produces experimental neurosis, should be apparent. We might therefore venture the prediction that tropical species are, in fact, more prone to experimentally induced neurosis than are nontropical species, this being a concomitant of the greater stereotypy of their behavior and the greater stability of the environment in which they have evolved. Presumably, only those tropical species that are resistant to the neuroses provoked by sudden, often unpredictable, changes in their environment were able to move out of the tropics to exploit and colonize northern and southern areas.

Further, if our various assumptions about the behavioral capacities of tropical species have been correct, we might predict that when tropical species are introduced into an area devoid of competitors, an area in which it would be possible for them to expand the size of their niches, they would be less prone to expand their niches than would be species from temperate regions. Varying environmental conditions have selected for an enhanced adaptability among nontropical species, an adaptability that is lacking in species of the tropics. This fact can be demonstrated under seminatural conditions by experiments involving the introduction of animals to areas previously uninhabited by them. The kind of observations that bear upon this point are illustrated by those noted by Crowell (1961), who has investigated the habits of the small number of birds found on Bermuda. Here he finds that the three commonest species of birds are also represented on the mainland. While on the mainland these three species share their habitat with a great number of other species, on the island of Bermuda they have most of the available space to themselves. In this instance, the birds in question do not appear to alter the dimensions of their niches in a radical way. Their niches have been enlarged, but not radically enlarged. The Bermudian representatives of the mainland species do not feed in radically different ways, although in the absence of competitors they might easily adopt such new ways. They do not nest in radically new places, although, again, in the absence of competitors they might choose new kinds of nesting sites. The size and shape of their territories do not undergo any kind of radical change. In short, Crowell finds that the Bermudian species do very much the same sorts of things on Bermuda that they do on the mainland, though there has been a broadening in their repertoire or the range of microhabitats in which these animals exhibit their normal behavior. When tested in the laboratory with artificial foliage (see below for details), there was some indication that the Bermudian birds were less stereotyped in their preferences than their mainland conspecifics (Sheppard et al., 1968). Van Valen (1965) and Grant (1968) had found a similar broadening of preferences or behavior among insular species they studied. In this instance, of course, as we are dealing with birds that are not primarily tropical in nature, their behavior confounds our prediction, which would have been that new dimensions would be added to their niches. This confounding could occur because our initial hypotheses are incorrect, because the conclusions drawn from the hypotheses are incorrect, or, alternatively, because historical or ecological factors have, in this instance, worked to mitigate the effects that would be predicted on the basis of behavior alone. Whatever the case, it is clear that there is a keen necessity for further analyses of this sort, as well as laboratory examinations of tropical and north temperate species that are designed to establish the readiness of these birds to expand the dimensions of their own niche.

A number of experimental attempts to measure directly the stability of habitat preference of various birds have been made in the hope of being able to make comparisons between tropical- and temperate-zone passerines and nonpasserines. As this work is still in an early stage, few comparative data are yet available. Most of the work pertains to the chipping sparrow (*Spizella passerina*) (Klopfer, 1963, 1965; Klopfer and Hailman 1965).

The purpose of the experiments in question was to determine the effect of early experience with certain foliage characteristics on future foliage preferences. Relative measures of the stability of particular preferences were also made. Altogether, four groups of chipping sparrows were assembled for this purpose. The first and second groups consisted of adults trapped on their usual breeding grounds. The third and fourth groups consisted of hand-reared young that were removed from their nest no later than the twenty-fourth hour after the opening of their eyes. Groups 1 and 2 were housed in large outdoor flights, supplied with both pine and broadleaf foliage, while the hand-reared birds were retained indoors in a small, cloth-covered cage. Group 3 saw no foliage whatsoever during rearing; group 4 was in constant contact with oak twigs and leaves.

The test chamber, within whose confines foliage preferences were examined, consisted of a room 12 feet long, 8 feet wide, and 8 feet high. A light gradient was established along the long axis of the room by means of a series of fluorescent fixtures attached to the walls, ceiling, and floor. The intensity of light at the various perches ranged from a maximum of 500 footcandles to a minimum somewhat below 0.8 footcandles. As one of the lights was placed in the center of the floor, facing toward the back of the room, the intensity of the light did not decline uniformly from back to front.

Rather, at the center, there was an abrupt decrease in intensity, so that it was possible to speak of a "light" and a "dark" half of the room.

Along two opposite sides of the room were arrayed steel racks consisting of two parallel sets of bars stacked ten inches above each other and running the length of the room. These assured the provision of adequate and uniform perch opportunities. It was to these bars that twigs of pine (on one side) and oak (on the other) were secured. Since food and water were available only in the center of the room and since position and temperature effects were randomized by alternating the positions of the two foliage types, it was thus possible to reduce relevant variables to two: foliage type (which might include relative size, shape, contours, or so forth) and light intensity.

Birds were released singly into the chamber, first for a 48-hour habituation period and then for several days of observation. These observations were made from behind a darkened screen of military camouflage cloth, which provided a window at the dark end of the chamber. The behavior of the bird, which included clinging to the screen inches from the observer's face, indicated that subject and experimenter were adequately separated.

By means of a panel of microswitches that activated timers and event counters, the observer could record the duration and frequency of visits to various parts of the chamber. These observations were randomized as to the time of day so that all portions of the animal's activity cycle were sampled.

Finally, the observations were repeated with the birds in outdoor flight cages (24 by 8 by 8 feet), each of which contained one small broadleaf and one small pine tree.

The preferences of the chipping sparrows trapped as adults were clearly for the pine foliage when they were examined in the test chamber. Only one bird out of ten spend less than 50 percent of the observation period in the pine. In the outdoor avaiary, these same birds chose pine and broadleaf about equally. It should here be emphasized that young pines provide far fewer perches per unit height or volume than most broadleafed trees. No preference for any particular light intensity was evident, in contrast to wild-trapped whitethroats (*Zonotrichia albicollis*). The latter made no distinction between the foliage types but habitually settled in the darker portions of the chamber.

The second group of wild-trapped adults was introduced into a test chamber in which the quantity of pine-decorated perches had been reduced by one-half. The consequence was that six of the ten birds now shifted the bulk of their activities to the oak.

The isolates reared without sight of any foliage, when tested at ages of two to four months, were indistinguishable in their foliage preferences from their wild conspecifics. The isolates reared with oak, however, showed an increased preference for oak foliage, half of the eight birds tested choosing oak more often than pine. The mean time spent in pine by all the groups was also proportionately reduced. Retests several months later, during which period the birds had been placed in outdoor flights supplied only with broadleafed trees (though with pines visible at a distance), demonstrated considerable stability in the preferences of any individual. Subsequently, the birds were removed to an aviary supplied only with pine foliage. It will be of interest to note whether an extended period with pine will reverse the preference established by experiences as juveniles.

Several conclusions of interest may be drawn from these simple experiments. First, while the chipping sparrow is clearly opportunistic, using whatever trees provide the most suitable perching sites, it does have a preference for certain foliage type that is independent of early experience. (It must be noted that the hand-reared birds came from several different nests whose positions with respect to height, tree species, and foliage density varied considerably.) Given a particular experience, in this instance a protracted exposure to oak leaves, the normal preference can be overcome. Whether this newly established preference is irreversible has yet to be demonstrated. In the meantime we can conclude that the behavior of at least this one species of north temperate zone passerine accords well with the predictions of Klopfer and MacArthur (1961) about what one would expect of a bird equipped to deal with a varied and varying environment. The plasticity of the species appears to exist on both a phenotypic and a genetic level, as witnessed by the small number of aberrant individuals in each group whose preferences deviated strikingly from its companions. A repetition of these same tests, using representative tropical species, allows a closer scrutiny of the hypothesis that tropical species, by and large, are less opportunistic and more stereotyped in their behavior. This behavior, it will be recalled, is assumed to mean that their niches are smaller. Only a few individuals have thus far been examined, but nine tropical tanagers (family Thraupidae) did show some interesting contrasts to the chipping sparrow. Among hand-reared foliage-deprived birds, the sparrows were more stereotyped in their visual preferences than the tanagers. On the other hand, a particular visual experience served to constrain the tanagers but not the sparrows. Under normal rearing conditions, then, the tropical tanagers would be expected to show greater stereotypy (Klopfer, 1965). However, the additional studies seemed to indicate that the tropical species did retain a considerable potential for expanding their repertoire. It begins to appear that their greater stereotypy may as likely be a result as a cause of increased species diversity.

Finally, we may consider differences in the variance and diversity of organ size, for this factor is of considerable relevance to our argument. If tropical species are more specialized and stereotyped in their behavior, the organs that are most directly concerned with this specialized behavior should show a reduced variance as compared with those of a jack-of-all-trades. In birds it makes sense to consider bill length as an indicator of specialization in food habits (not necessarily the best indicator, though clearly the most easily measured). Let us take the bill lengths of an imaginary group of birds of different species, all of which utilize a wide range of foods, and compare the distribution of their bill lengths with that of a group of rather specialized species. The total range of lengths would be expected to be similar in the two cases. The nonspecialized species, however, should have a rather low diversity; that is, most species should have bill lengths close to the mean for the entire group. With increased specialization, a greater diversity should result, though the mean might remain constant; that is, the different bill length classes would have a more nearly equal number of species representing them.

In Fig. 3–5 are graphs based on the mean bill lengths of adult males (exclusive of aquatic species) of all species recorded from Panama and Massachusetts. It must be added that sample sizes varied from one species to the next. Since the data were culled from a variety of sources, there are undoubted differences in the reliability of the figures for different species or regions. However, bearing these qualifications in mind, it will be seen that our prediction is not altogether wrong. The nonpasserines in both tropical or temperate latitudes are characterized by a greater diversity than the passerines. Nonpasserine individuals make up a greater proportion of the total

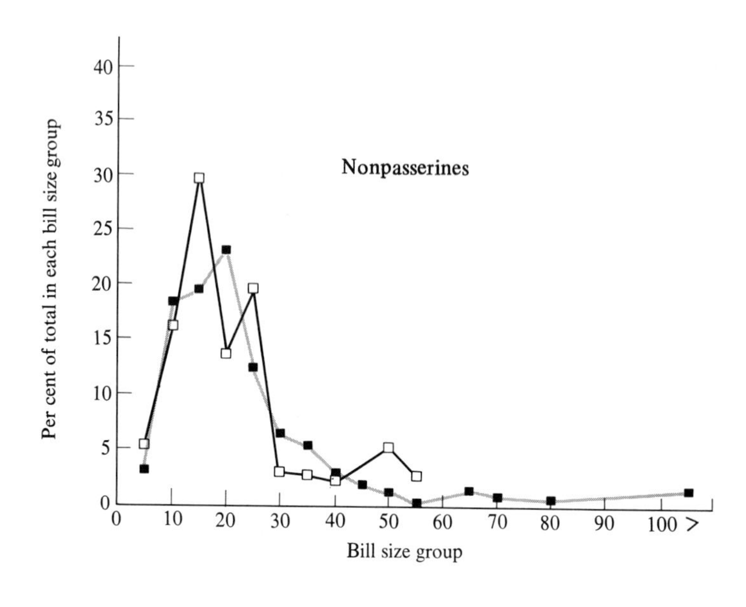

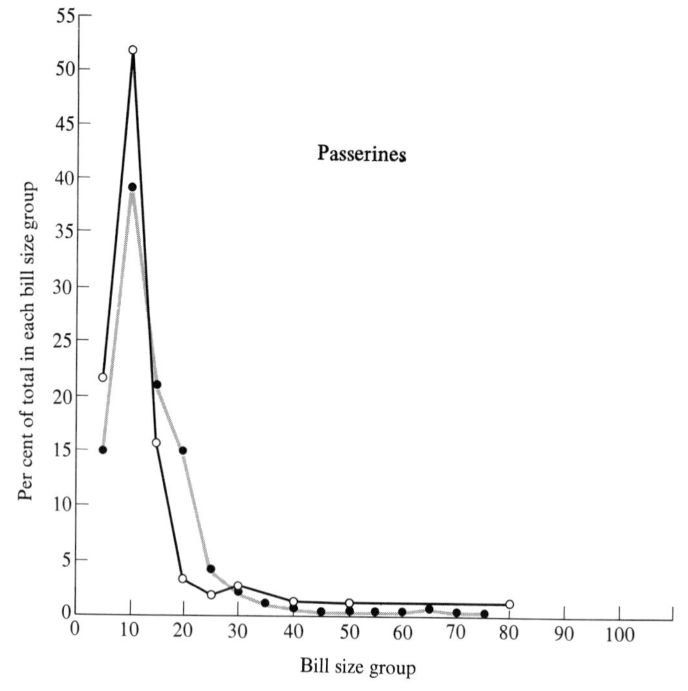

Fig. 3–5 These four graphs illustrate the greater adherence of passerines (as compared with nonpasserines) and of temperate-zone birds (as compared with tropical birds) to the jack-of-all-trades principle. Passerines are represented

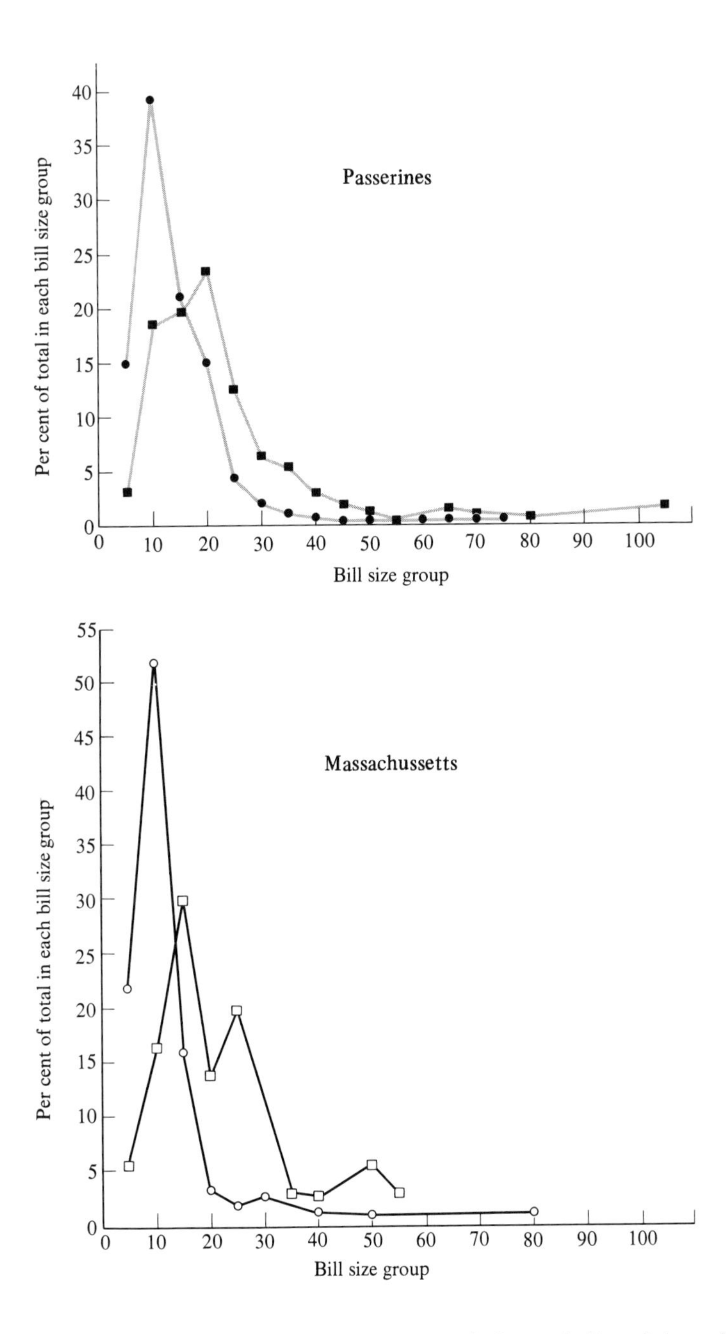

by circles and nonpasserines by squares. Tropical are indicated by solid figures and temperate-zone species by open figures. The broader and lower the curve, the greater is the number of specialist species.

avifauna in the tropics than farther north, showing that a greater bill size diversity is indeed found among tropical avifaunas. Further studies along these lines, considering both the Old World and the southern hemisphere, would thus seem quite advantageous.

Finally, the various studies of Schoener (1965, 1968), Grant (1968), and Pulliam (1970), all of which bear on these points, even though not necessarily strengthening the particular argument advanced here, should be noted. They raise the further question of differences in genetic variance among population of different size, an argument of obvious relevance if areas of high diversity also have fewer individuals of any one species. High genetic variability (which requires large populations) can be particularly important under changing environmental conditions, especially for short-lived organisms such as insects. *Drosophila* populations, for instance, with the greater degree of heterozygosity showed the greater fitness in unstable environments (Ayala, 1968). This may be the reason why temperate areas cannot be so diverse in their numbers of species.

(A recent monograph that has just come to my attention indicates why it is likely that "in attempting to explain gradients in bird species diversity, one must consider not only a wide range of ecological parameters but also the geologic history of the region in question and the distributional and evolutionary history of the species under study" (Howell, 1971, p. 235). I do agree that I have erred in the direction of oversimplification in emphasizing a single explanation for a phenomenon that increasingly appears to depend on a multiplicity of causes.)

FOUR

How Are Species Kept Distinct?

In Chapter Three we considered some of the possible explanations for the enhancement of the tropical fauna. We rather took for granted the fact that diversity is a characteristic of all regions of the globe. That is, we usually find more than one species, often many more than one species, occupying a particular habitat. The multiplicity of species is presumed to be the ultimate outcome of a competition for resources that results in progressive specialization (Gause, 1934; Hardin, 1960). The immediate mechanism underlying species formation has generally been thought to be geographic separation of sufficient duration to allow the accumulation of genetic differences (Mayr, 1963). A geographically continuous population should nonetheless be capable of fracturing into distinct species if it is polymorphic and inhabits a heterogeneous environment (Maynard-Smith, 1966). This issue as to whether sympatric speciation can occur is complicated by claims of Ehrlich and Raven (1969) that species are not evolutionary units with a common gene pool. They argue that gene flow is far more restricted than commonly believed: Many local assemblages of a so-called species exchange no more genetic material inter se than they do with members of other species. Whether or not morphological or behavioral differentiation occurs depends wholly on how similar or different local selective forces are.

Whatever the mechanisms whereby species arise, and this remains a contentious issue, it is at least clear that barriers preventing hybridization will be preserved. Hybridization, after all, would reduce the effectiveness of niche diversification, thereby intensifying competition. A reduction in species diversity makes for a reduced tolerance of environmental vagaries, too (MacArthur, 1955). Foxes (*Vulpes* sp.) that are part of a community that includes only one species of bird and rodent would have a slimmer margin for survival than those in communities with several species (also note Paine, 1966, 1969).

The optimum number of coexisting species will be a function of several conflicting forces. As indicated in Chapter Three, there is reason to believe that in the less stable habitats a premium will be placed on the jack-of-all-trades. Under more stable conditions, the specialist will be favored. The latter can proliferate into a larger number of separate species without increasing the severity of interspecific competition than can the former. At the same time, random environmental fluctuations will cause the periodic extinction of specialist species (whenever their particular requirements can no longer be met), the more so since the number of individuals per species must be less when the total number of individuals the habitat can support is divided among a greater number of species. These considerations, however, are irrelevant to our present purposes, as here we wish to examine the manner in which boundaries between species are maintained. We have seen that these boundaries are favored by natural selection because they allow a more efficient utilization of resources and the appearance and coexistence of more individuals. But the question of how this diversity is continually maintained is a problem that is worthy of considerable attention. It is useful to group our answers into two categories. The first, *built-in constraints*, refers to limitations preventing hybridization of a sort that an individual organism has no way of overriding. Its "hardware" constrains its behavior. The second category, *acquired constraints,* corresponds to "software" or programmed constraints. These are built up during the lifetime of the individual, as a result of specific experiences, usually gradually and usually subsequent to birth or hatching. The operational basis for the distinction is simply that the hardware is able to develop in the absence of conspecifics, while the software is not. This is obviously an entirely arbitrary criterion, but since nothing further is implied by the dichotomy, this does not matter. Certainly, there is no reason to consider the hardware–software distinction as in any way corresponding to the old and sterile instinct–learning dichotomy. That pair of terms implied a fundamental difference in underlying mechanisms. All that is implied here is a difference in the plasticity of the response to environmental inputs at different stages of development. As Kuo (1960) has pointed out, the behavior of an organism at any instant in time is a function of its developmental history and present environmental setting, as well as its biochemical and neuroanatomic features:

> It goes without saying that behavior is predictable only when the essential factors in all of these three categories are well known and understood. And yet, in our behavior studies in the past few decades the developmental aspects were most often neglected. The Watsonian behaviorists and their allies as well as the Pavlovian conditioned reflexologists have been contented with the oversimplified SR formula. The Gestalt school and more recently the ethologists have laid great stress mostly on the present environmental settings. But the neglect of developmental history has made a great many of our behavior studies look fragmentary and void of fundamental understanding of animal behavior. And it is, partly at least, due to such a neglect that a number of investigators of animal behavior have to resort to such concepts as inheritance, maturation, or to introspective or even anthropomorphic concepts in order to explain their experimental results. No one can refute the view that every bit of behavior is resolvable into physiological factors, neurological and anatomical, or otherwise. However, every animal is the creature of its developmental history and reacts to a specific stimulus in a specific environmental setting. [Pp. 238–39.]

To this we would add that the development of every creature is a continuous and continuing process. There is no logical reason to consider that certain kinds of developmental processes cease at the moment of birth or hatching, others commencing then. From what we know of embryogenesis it is clear that the organization of the central nervous system has its beginning early in the period in which the ectodermal placode that will later form the nerve cord and the central nervous system is formed. The organizational changes that occur then continue to one degree or another through maturity and death. Whether these changes involve alterations in the size of synaptic end bulbs or merely reorganization in patterns of conduction is not of any great importance. The fact remains that to one degree or another any complex series of acts is ultimately dependent on some kind of organized neural substrate and that the development of this substrate occurs as continuous process. Thus, while it is possible to speak of instinctive behavior in terms of a certain operation that one performs or in terms of certain conditions under which particular reactions may occur, it does not make sense to speak of instinctive behavior as a functionally or neurologically distinct category. To quote Lehrman (1956):

> We consider attempts to find discrete central cores of behavior, pre-existent innate components of behavior and autogenous central sources of motivation to be restrictive in the sense that they exclude much of the relevant physiology, and unrealistic in that they depend upon undemonstrated forms of energy and of anatomical organization [P. 503.]

In terms of operational criteria, then, it may make good sense to categorize behavior as being largely due to hardware or as being instinctive, that is, stereotyped and resistant to change, or learned, that is, flexible, plastic, susceptible to change, and acquired gradually. Needless to say, other definitions or distinctions between these two among many other categories of behavior could be made. But it must always be remembered that these are distinctions that are made on operational criteria for purposes of convenience, and there is no valid basis for assuming that they reflect any fundamental differences in neural action or organization.

By way of illustrating the interaction of experiential and anatomical factors that shape individual behavior, let us examine the development of a particular animal. We shall select gulls (*Larus*) as our subjects, both because of the wealth of comparative data already available on these species and because the development of the young proceeds through distinct and readily definable stages. This is not to say that complete behavioral dossiers exist for any one species. Indeed, much of what follows is hypothetical. Nonetheless, it should provide us with some concrete examples of the interactions referred to above, specifically with regard to the ontogeny of visual recognition.

THE INTERACTION OF THE INNATE AND THE ACQUIRED*

An important aspect of the adaptive features shaped by natural selection is the ability of organisms to unite selectively with other individuals of the same species. This ability is obviously of particular significance for reproduction, but it is no less important in the search for food or in responses toward predators. But just how are conspecifics recognized? This is not known for a single vertebrate species. Gulls, however, have been studied in so thorough a fashion by ethologists that one can develop a provisional picture of the process by which their recognition of conspecifics may develop.

We are really not certain whether recognition of conspecifics is a unitary process. An intuitive belief leads to the assumption that a laughing gull (*Larus atricilla*) knows that it belongs to a particular group of similar individuals irrespective of what their behavior at that moment is. An alternative belief would be that a particular gull in any functional state reacts selectively to particular social stimuli and that under natural conditions its conspecifics are the animals that optimally provide those stimuli (note the *Kumpan* concept in Lorenz, 1935).

Actually there are indications that a gull doesn't always recognize its own kind. Gulls of different kinds are found together in loose flocks in winter, seeking food together, and at this time they don't seem to show any kind of recognition of conspecifics. Other kinds of birds hold themselves apart from these mixed flocks. Ducks, for example, build mixed flocks among themselves but stay away from the gulls. Even the rather more closely related terns stay away. Thus, in winter time, gulls may have some kind of a self-recognition

* This section is a translation from the German by P.H.K. of an essay originally prepared by J. P. Hailman and translated by W. M. Schleidt into German.

system operating, but it seems to be less selective than in spring and allows inclusion of other species of the same genus or family as well.

Sometimes the capacity for recognition of "own kind" is greater than at first appears to be the case. If one watches a mixed flock closely, one feeding on a sand bank, one might note, for example, that one or two ring-billed gulls (*Larus delawarensis*) fly up, followed by their conspecifics, while gulls of other species remain quietly behind. In short, there are situations in which gulls actually do seem to treat congenerics as their social companions and yet other situations where only conspecifics serve.

N. Smith (1966) has shown that morphologically similar arctic gulls of certain species distinguish their kind on the basis of the color contrast of the eye ring and iris. By painting over the eye ring, Smith was able to induce a variety of species to hybridize. Thus a single morphological characteristic is all that is necessary for recognition of conspecifics. These same species form mixed flocks in winter. Unfortunately, we don't know whether there is still a preference for own kind in winter, but if so, it has to depend on some characteristic other than the eye ring. This is present only during the breeding period. Perhaps the voice or the coloring of the wings is used.

In short, if gulls choose their conspecifics in different situations according to different social stimuli, then one cannot consider the problem of recognition of own kind as a single or unitary process.

From the standpoint of evolution, it hardly makes any difference how gulls distinguish their conspecifics; it is only important that they do so. The "how" remains of interest, however. Thus it is worthwhile to examine some studies on the recognition of gull parents by their chicks for the insight it provides into the development of what appear to be "innate" preferences.

The newly hatched gull chick apparently has only a very limited ability to distinguish its parents from other animals or objects in its environment. This ability does suffice to direct its begging response to its parents and thus to elicit feeding by the parents. Feeding acts as a reenforcement, in a psychological sense, for the learning of the characteristics of the parents. Eventually, the chicks recognize their parents on a basis of some kind of Gestalt perception. The relationship between parent and young, strengthened in this manner, helps the young to recognize conspecifics after it reaches the age of fledging, through a process of stimulus generalization and learning transfer. Thereby it can integrate itself in the social group. The following findings provide support.

In most species of gulls the newly hatched chick pecks at the bill tip of the adult, which is contrastingly colored, and thereby releases regurgitation (by the adult) of partially digested food. This food is usually vomited onto the ground, and if the chick doesn't pick it up, the parent picks it up again. Then the chick takes it from the parental bill. Older chicks often feed directly from the ground or take the food out of the open bill of the parents the moment that it is regurgitated.

When chicks see food for the first time they don't recognize it as such. They do, however, peck at everything within reach of their bills and eventually, through trial and error, encounter the first particles of food. Or, by pecking at the bill tips of older siblings while these are already feeding, they make contact with food (Hailman, 1961, 1967). Once a particle of food is in the bill, it is swallowed quickly. One particle of food already suffices to teach the chick something about the optical characteristics of the food (Hailman, 1967). This shows that the newly hatched gull chick is capable of very rapid associative learning.

Although the newly hatched gull chick directs a certain proportion of its pecking movements to different objects in the nest and in its immediate surroundings, most of its pecking is directed to the parental bills. If eggs of the laughing gull are artificially incubated and the chicks hatch in the dark, so that all possible influences of the environment on the optical capabilities of the chick are excluded, then one can determine what stimuli will maximally release pecking. Experiments with models of bills showed that the following characteristics were most effective in eliciting pecking: (1) object darker than its background, (2) perpendicular orientation of the main axis, (3) a particular width (about nine millimeters), (4) movement in the horizontal plane and perpendicular to the long axis, (5) a particular speed of movement, and (6) red or blue objects on a yellow or green background (Hailman, 1966, 1967, and other unpublished experiments). In addition, we now know that the strength of the illumination and dark-light contrast of the background influences the frequency of pecking in a complex manner, although all the details are not yet sufficiently clear. The three-dimensional projection of the stimulus object also seems to be of significance.

Despite the length of this list of characteristics, the actual form they add up to is absurdly simple. Of course, it has to be simple since we were able to explain it and analyze it into its components.

Tinbergen and Perdeck (1950) have carried out similar studies, as have Weidmann (1959, 1961), Weidmann and Weidmann (1958), and Collias and Collias (1957), and they came to similar conclusions. In all these cases the corresponding preferences of the newly hatched chicks corresponded to the morphological and behavioral characteristics of the parents with but one conspicuous exception. In the herring gull (*Larus argentatus*), the bill of the parents is roughly double the width of the chick's preferred width. While the other species studied have red bill tips, that of the herring gull is yellow and has only a red spot on the lower mandible. The diameter of this spot, however, is just about the optimal width insofar as the chick's preferences are concerned. This is an instance where the exception proves the rule.

If one makes measurements of the bills of the roughly 44 species of gulls and compares them, one finds that in the case of large-billed species the red on the bill is reduced to a spot on the tip or on the lower mandible. (This is at least true for the 25 species thus far studied.) This argues strongly for the fact that the preference of chicks for red is phylogenetically older than the red on the bill of the parent.

Certain species of the genus *Larus* (according to the revision of Moynihan, 1959) have no red on the bill. The kittiwake (*Rissa tridactyla*) has a pure yellow bill. The chicks of this species take food from within the bill of the parents, and this bill is colored red inside (Cullen, 1957). The ring-billed gull only has a black ring near the tip of the bill, but despite this the chicks peck in the same manner as those of the other species mentioned (John Emlen, personal communication). The black ring provides a light–dark contrast that is itself an important characteristic. Moreover, this particular species has a red lining to the bill, just as does the kittiwake. *Larus bulleri* has a completely black bill somewhat of the size of that of the laughing gull. In this species, the newly hatched chicks preferably peck black instead of red models (C. Beer, personal communication). Finally, *Larus furcatus* on the Galapagos Islands has a black bill with a white tip. This species is night-active (Hailman, 1964b). The white tip functions as a particularly good stimulus under low light intensities. *Larus furcatus* reacts less toward movements than, for example, does the laughing gull and begs intensively even under very low light intensities, while the laughing gull remains passive under similar light conditions.

According to these comparative studies, the perceptual preferences of newly hatched chicks of all gulls thus far tested seem to be roughly similar. They have a built-in discriminative capability that assures that in the first instance they will react to their parents and not to stalks of grass, nest material, or other objects in their immediate environment. This preference could be coded in the eyes through relatively simple mechanisms (Hailman, 1964a, 1966, 1967), a discussion of which would, however, carry us too far afield here. In any case, these discriminative mechanisms are not sufficiently fine-grained to assure a specific recognition of conspecifics. This ability must develop subsequently.

An example that seems to contradict the rule of relatively simple perceptual mechanisms in newly hatched gull chicks was published by Tinbergen and Perdeck (1950). Herring gull chicks pecked more frequently at a model that had a red spot on the bill in its natural place, that is, on the lower mandible, than they did to a similar model that had the red spot on the forehead. The interpretation of this result by N. Tinbergen (1951) was as follows: The chick has a configurational image of the head of the parent, a Gestalt, that is destroyed or disturbed through changes in the spatial relationships of the components. More recent experiments on herring gulls (Hailman, 1967), however, have shown that the differences between the two models that Tinbergen and Perdeck used were not merely in the differing position of the red spot but also in the simultaneous change in three other variables. If (1) the height of the red spot relative to the eye of the chicks, (2) the speed of movement of the spot, and (3) the length of the arc in which the spot is moved are held constant, then the position of the red spot relative to the contours of the model does not produce a difference in the reactions of the chick. In other words, the chicks do react to the simple characteristic, red spot, and not to the configuration of the entire head.

The fact that such simple preferences do not suffice to provide for a discrimination between different species could be shown through direct experiments with models of herring gulls and laughing gulls using chicks of both species. In both species the chicks peck equally frequently at the models of their own and the other species (Hailman, 1967).

The significance of the perceptual mechanism of the chick is not the provision of a discrimination between species but only the assurance of feeding and the development of a relationship with the parents. This makes it possible for the chick to discriminate between its parents and the ubiquitous predator. There is no particular need for own-kind discrimination at the time of hatching, and natural selection has apparently not developed any particular mechanism to accomplish this.

When chicks are older, one or two weeks, a successful model has to resemble the parents very closely, even to details. Older chicks of the laughing gull react only to models that have the correct form and proportions of head and bill, while the newly hatched ones, as noted, will peck at very simple models, even ones lacking a head. One-to-two-week-old chicks no longer peck at models of the species other than their own. They now can distinguish between their own kind and other gulls (Hailman, 1962, 1967).

This type of recognition of conspecifics is probably dependent on a learning process in which food serves as reenforcement. When chicks of the laughing gull are fed with the herring gull model, later on they will no longer react to the model of the laughing gull. Control animals that have been fed by hand fail altogether to distinguish between models of the two species. This process has been termed *perceptual sharpening*. It is related to the classic conditioning of Pavlov, except that here the unconditional and the conditional stimuli originate from the same object, the head of the parent (see further details in Hailman, 1967).

One can imagine that the hungry chick, which sees its parents flying toward it, ultimately learns all the essential characteristics of the parental body. Whether the chicks thereby also learn the specific features of their individual parents is still uncertain. Herring gulls do seem to recognize their own young after about five days (N. Tinbergen, 1953). Whether or not and how long it takes for the young to recognize their own parents is still unknown.

Before the young gulls fly for the first time, they wander about the breeding colony. During this wander phase the recognition of the parents could be learned, or at least reenforced, due to the fact that, in some species, breeding adults defend a territory not only against adults but also against strange young. They lunge and peck at wandering chicks and may injure or even kill them. Such attacks from the hostile adults of the same species probably help the chick to discriminate them from their protecting and food-providing parents, at least if the attacks are survived.

In a few species, the parents continue to feed the young even after they have been fledged for a certain time. Begging then no longer consists of pecking but in a short upward shaking of the horizontally held bill, while at the same time a high tone is emitted (bill tilting; Hailman, 1967). Later, a third begging movement, head pumping, appears. The continuing dependence of the young on food from the parents and the effectiveness of begging in the elicitation of feeding have as an effect the maintenance of a strong bond between the young and its parents even after flight has been attained.

We don't yet know how specific such bonds are. Franklin's gulls (*Larus pipixcan*) collect chicks and feed all they can find. Up to ten have been found in a single nest (Roberts, cited by Bent, 1921). This unusual behavior seems to be a particular adaptation that prevents the chicks from perishing when they accidentally go too far from the floating nest upon which this species breeds. Foster parents also seem to exist in *Larus bulleri*. Here the chicks have a personal bond with one parent and not with the nest (C. Beer, personal communication). In such a case the original bond with the true genetic parents is certainly extinguished. In herring gulls, the chicks appear to beg for food from their parents only in their own territories (N. Tinbergen, 1953). According to Strong (1914), among the herring gulls the chicks are fed for six weeks, and sometimes half a dozen will be fed by a single adult. The normal brood size of the species is three eggs. He suggests, however, that the parents regurgitate only to their own young and that the additional young merely exploit this to their advantage. A hand-reared herring gull at the Drumlin Farm Sanctuary in Massachusetts begged from people throughout its first winter while it seemed not to pay any attention to other gulls and other birds on the pond. There was also observed a young ring-billed gull that followed the adults until winter time in order to beg food from them. These findings argue for the fact that the characteristics of the species are learned prior to the conclusion of fledging.

The strong social bonds between fledged young and adults are probably of great significance for further learning. This is particularly noticeable in those cases in which the adults acquaint the young with unusual feeding areas, as apparently happens at the conclusion of the molt. In this way the development of species traditions commences (Hailman, 1960). In such instances the young not only have the opportunity to learn traditions but also particular optimal characteristics and calls of the conspecifics during their search for food. The characteristic folding of the wings and the "falling" or diving posture of a gull that is being fed by the passengers of a passing ship draw other gulls into the area from a distance as far as a mile. Gulls that simply follow a ship without making the falling movements don't have this effect (personal observations on herring gulls, ring-billed gulls, and laughing gulls). Other characteristic visual stimuli or acoustic signals are also used by various types of gulls and associated with social feeding (Frings et al., 1955), and youngsters could just as well be conditioned to them. Subsequent to fledging it is further possible to condition reactions toward conspecifics by means of secondary cues.

If we summarize our present understanding of the recognition of conspecifics in gulls, we develop the following picture. Newly hatched gulls preferentially peck models of a particular and simple design. In most species, the color of the bill is such as maximize pecking. Inexperienced chicks are rewarded with food for pecking the parent's bill, and in this way they learn further complex characteristics of the bill and the head. Through secondary conditioning they finally learn to recognize the adult as such. Other kinds of treatment associated with parental care, for example, brooding, also serve as reenforcements. Begging, which replaces pecking, is directed toward the parent, whose characteristics have now been learned. Inasmuch as the young follow their parents in their first fall of life they have the opportunity to learn further characteristics, particularly those dealing with situations concerned with food finding.

It is difficult to imagine that the mechanisms of recognition of conspecifics, which involve complex *Gestalts*, can be programmed without the developing sensory systems and the central nervous system being overburdened, since they are particularly susceptible to disturbance. The recognition of conspecifics could be safely learned, however, by progressively building upon a few simple preferences and responses.

THE ACTION OF BUILT-IN CONSTRAINTS

Having now considered the interactions of acquired and built-in constraints, let us consider them separately.

Constraints that limit the sexual partner of an individual to a particular species can be constructed in any one of a variety of ways. One may, first of all, create a genetic incompatibility. This is the case where the number of chromosomes differs between the two species, or the shape or form or chemical makeup of the chromosomes differs so that normal synapsis cannot occur. There may be subtle cytochemical or similar differences that make for incompatibility and so prevent normal mitosis from taking place. Such constraints, while effective in preventing hybridization, are, however, not maximally efficient, particularly among forms that are capable of only a limited number of matings. Although their gametes will not combine to form viable zygotes, a mating has been wasted. Thus constraints of a grosser morphological sort should be expected to evolve in many of these cases, and, indeed, among a large number of forms there are gross morphological constraints that act as the primary barriers to the hybridization. The shape or position of the penis is rather variable in a good many species of the fly, *Drosophila,* for example, thereby effectively precluding hybridization. The form of the stigmata of plants may physically preclude fertilization by pollen of particular sizes or shapes, thereby limiting successful pollination to a very small range of pollen types. It is also possible for the barrier to be of a physiological sort that is dependent on differences in time of reproduction, either on an annual or on a diurnal basis. Thus, as was demonstrated earlier among a number of tropical tanagers (family Thraupidae), the peak of the breeding season of the various species may differ, some reaching their peak in May and others not until July. A diurnal variation may occur as well, which, indeed, has been demonstrated to be the case among a number of sympatric species of *Drosophila,* some of which show reproductive behavior only in the evening and others only during the day. Such differences in time of reproduction or of activity can then substitute quite effectively for either the morphological or genetic constraints that preclude gene interchange.

Yet another type of built-in constraint of a more strictly behavioral nature is dependent on the existence of what ethologists have termed *releasers* or sign-stimuli. The role of such sign-stimuli in preventing hybridization has been extensively reviewed by Hinde (1959). We are dealing here with the situation in which the stimulus to engage in reproductive behavior consists of a fairly restricted and specific form—movements, colors, sounds, smell, or a combination thereof—that normally emanate only from an individual of the same species and of the opposite sex. In the absence of this specific sign-stimulus, appropriate reproductive behavior cannot be released. Examples include particular species-characteristic colors, to which retinal cells are especially sensitive (Swihart, 1967); or courtship songs, which serve to segregate some sympatric species of birds (Lanyon, 1960; Stein, 1963); or even the temporal patterns of intromission (Diamond, 1970). These constraints represent a fairly inflexible form of behavior and can be as effective as the more overt morphological or genetic barriers.

THE ACTION OF ACQUIRED CONSTRAINTS

The category of acquired constraints poses problems that are of somewhat greater interest and that have been dealt with far less extensively than those subsumed under the head built-in constraints. When we talk about acquired constraints we are saying, essentially, that the organisms in question have to learn through their own experience the nature of the appropriate sex partner. How does this learning come about? There are situations in which animals, through an appropriate manipulation of their environment, can be made to attempt reproduction with an inappropriate organism. How does learning normally proceed to reduce or prevent such maladaptive attempts?

First, it is possible that for many species the relevant learning is of the type that has been called imprinting. When Lorenz (1970) first discussed imprinting, he assumed that one of the primary features of this process was that it did fix the reproductive responses of the mature organism to the model or surrogate to which it was exposed during its critical or sensitive period. The graylag geese (*Anser anser*) that followed him shortly after hatching were assumed by him to have their sexual responses imprinted onto him; classically, one of the primary characteristics of imprinting was this fact that a delay existed between the original exposure and the final manifestation of the response to the object to which exposure was made. Unfortunately, we know of no cases in which this claim can be considered to have been satisfactorily verified. In none of the instances that Lorenz and others have mentioned is the possibility of secondary reenforcement or conditioning after the critical period precluded. A graylag goose that follows an individual shortly after hatching but then continues to be fed by or played with by that individual during its juvenile period cannot be said to have had its sexual responses imprinted onto that human. The responses could as well have been conditioned during the juvenile period. The critical experiment would entail isolating the gosling from its human surrogate for the time from the end of the critical period until the commencement of reproductive maturity. The difficulty that such experiments pose in terms of the time and space required seems to have prevented most investigators of imprinting from concluding them. As a result, this particular feature of the definition of imprinting has often been dropped by investigators. One exception, which does support the notion of sexual imprinting, is a study of turkeys (*Meleagris* sp.) by Schein (1963), although another study of ducks (*Anas* sp.) by Schutz (1965) is ambiguous. The absence of more such studies is regrettable in view of their significance, though understandable in view of their difficulty. Their significance is underscored by a theoretical argument by Seiger (1967) defending the possibility of sympatric speciation through imprinting. Neither this aspect of imprinting nor the development of sexual preferences themselves are touched upon by those many studies of imprinting that focus on the "following response," which, at best, is only an indication that imprinting is taking place (Sluckin, 1965; Bateson, 1966; Klopfer, 1971a). In short, we must agree with Lorenz that imprinting (and by this we include the notion of sexual responses being fixed by the early exposure) may very possibly occur. The convincing experimental demonstration of this fact, however, has yet to be made.

Second, the recognition of the appropriate sex partner may be the consequence of a protracted relationship with the parents that allows for a habituation to the parent type. When sexual maturity is attained and aggressive behavior reaches a peak, it may be that only those forms to which habituation has occurred can be tolerated for sexual purposes. The important processes to focus on in such instances are those that assure a continuing and close relationship between mother and young.

Collias (1956) and others have uncovered some evidence that suggests that imprinting-like phenomena may play a role in the recognition of the appropriate sexual *Kumpan* even in such highly developed forms as sheep (*Ovis* sp.) and goats (*Capra* sp.). What Collias actually demonstrated was that the tendency of the female to accept her young was dependent on there being olfactory and tactile contact with the young for at least a few minutes directly upon parturition. A separation prior to such contact, even if of a duration as short as one hour and sometimes less, was often sufficient to lead to a rejection of that young by its mother. We do not yet know whether the reciprocal effect upon the young does exist, although Howard Liddell (personal communication) claims that his "rejected" lambs and kids show abnormal growth patterns and have an appallingly high death rate. A reciprocal effect upon the young is a possibility that clearly must be investigated. It is also now clear that the situation described by Collias and Collias is complicated by differences in the behavior of the various breeds or strains of sheep and goats as well as by differences in the duration of the separation that can be tolerated among primiparous and multiparous females.

In the first of their experiments on this problem, Donald K. Adams and Klopfer (unpublished data and Klopfer, 1971) suggest that for primiparous animals a separation directly following parturition is less likely to lead to the subsequent rejection of the young than is the case with experienced multiparous females. Perhaps this difference occurs because the afference at the time of birth is far greater for females who are giving birth for the first time. This is also the case with women of the human species, and on anatomical grounds there is every reason to believe that it is the same for sheep and goats or any other mammal. The stretching of the cervix and the vaginal orifice produces much more intense stimulation when the first infant is born than subsequently, and thus the reenforcement stemming from the birth-associated stimuli is far more intense in the first than in subsequent births. Perhaps the greater intensity of stimulation is what produces a stronger maternal–filial bond in primiparous animals and thus allows for a more protracted separation without the effect of subsequent rejection. There is evidence that cervical dilatation induces the release of oxytocin, which has been implicated in the induction of maternal responsiveness (Klopfer and Klopfer, 1968; Hemmes, 1969). At the moment, of course, this information is largely conjectural, but it does point to an interesting problem for research-minded psychiatrists and human biologists. Psychoanalytic data must surely be available on series of children of one mother, with reference to their birth order. Especially where these children are identical twins or members of a long series, their records would be of considerable value for study of the connection between maternal–young relations and the intensity of birth-associated stimulation.

Can we predict the circumstances under which one or the other of the mechanisms for assuring appropriate species recognition will be operative? Imprinting in its simplest form, where the initial exposure is attended by the following response, seems clearly dependent on a high degree of motor precocity. Among forms where a high degree of physical helplessness is the rule, as in young songbirds, there is a physically enforced association with the parents, and very rapid learning processes are not essential. Since the rapid fixation of responses in imprinting can lead to the development of maladaptive fixations as readily as to adaptive ones, one would not expect imprinting to be characteristic of any forms other than those capable of actually dispersing from their parents shortly after birth or hatching. Further, many birds, especially in the tropics, form mixed flocks within which there is little or no species segregation. Obviously, the processes of habituation must be limited to the nestling period if the young of such species are to learn to discriminate their own from alien species. Alternatively, in such forms built-in constraints may be relatively more important than in the typical single-species flocks of temperate zone locales.

In individuals with a relatively limited central nervous system, built-in constraints, being most economical of nervous tissue, would be the ones that natural selection would be first to favor. On the other hand, in organisms whose motor system develops at a relatively more rapid rate than their sensory systems, a rapid learning process such as imprinting becomes more advantageous, assuring that a maternal–filial bond is established before the young disperse. Presumably this is why imprinting is more commonly found among those birds that are most precocial than among the others. The data about imprinting and imprinting-like responses that we have thus far garnered all suggest this correlation between the prevalence of imprinting and precocity. There are, however, many instances where there is, so to speak, neural material to spare, and where there is a physically enforced dependence of the young on the parent. In these instances there is no necessity for natural selection favoring the development of constraints of the sort that we have discussed thus far. It is in these instances, then, that the recognition of the appropriate species and, of course, of the parent can be expected to develop gradually as the result of repeated reenforcement of associative learning.

We do know that among quite a range of organisms, and man is not least among them, one can find individuals whose sexual responses are most inappropriately directed toward other members of the same sex or members of the same or different sex but of a different species, genus, or even family. In some instances, indeed, sexual activities may even be directed regularly toward inanimate objects. (Among humans this occurrence has been termed fetishism.) Yet, despite the mass of clinical evidence available regarding the existence of these sexual misalliances, we know regrettably little about the processes that lead to their establishment. This lack of knowledge is not altogether surprising for such studies inevitably entail a great deal of time and attention to a mass of detail. Hayes and Hayes (1951), who attempted to bring up a young chimpanzee (*Pan* sp.) in their household as if it were a human child, provided themselves with the opportunity of discovering to what extent the sexual responses of this species are modifiable as a result of their upbringing. We do not know whether an animal reared as was Vicki, their foster chimp, would forever after direct its sexual responses to members of the species of the foster parent or whether built-in constraints exist that assure a recognition of the appropriate mate when the appropriate mate is available. This last statement needs perhaps to be emphasized. The fact that a sexual misalliance occurs cannot be interpreted as particularly significant if no opportunity for a normal alliance was provided. A sexually motivated animal deprived of its proper mate may well accept a substitute, but unless the proper mate is also presented as an alternative the acceptance of the substitute cannot be interpreted to mean that the animal's normal preference has been altered. What is needed, then, is an experiment, or a whole series of experiments, in which the young of different species are isolated from their normal parents for varying periods of time at varying stages in their development and upon sexual maturity are given an opportunity to express their preference for one of an array of sexual surrogates. By such means we may eventually find it possible to be a little more precise about the mechanisms that lead to the gradual development of a preference for a particular sexual object.

Effects of Domestication

Among domestic animals, improperly directed sexual behavior seems especially common. In part this may be an artifact resulting from the fact that domestic species can be more readily observed than their wild counterparts, but at least in part it probably represents a true concomitant of domestication. Thus it is worth considering the relationship between the processes that have led to the domestication of certain animals and those that are involved in the selection of the conspecific mate.

Spurway (1955) has argued, most reasonably, that those animals most likely to have been domesticated were the animals who most readily could learn to transfer their responses from individuals of their own species to individuals of the human species. In other words, organisms that could be readily imprinted and thereby readily transformed psychologically would be those who would be most likely to undergo domestication. A contrary point of view was also advanced by Klopfer (1956), who argued that since, in a barnyard situation, the probability or the possibility of imprinting onto biologically inappropriate objects was so much greater than that obtaining in the wild and since such imprinting was likely to interfere with normal reproduction, there would tend to be a selection against the ability to be imprinted during domestication. Actually, it is possible to conceive of a situation in which both explanations apply, the initial domestication depending on the ease with which sexual and other social attentions can be transferred to the human species and the continued existence of the then domesticated species being dependent on the fact that the transfer of sexual responses to humans is not absolute.

In any event, to assess the validity of these two viewpoints, there has been at least one recent attempt to compare the degree to which imprinting may occur in domestic and wild races of the same species. This work was performed by Gottlieb (1961a), who made use of the Pekin duck, a highly domesticated strain of *Anas platyrhynchos*; domestic mallards of the same species that had not lost the wild-type plumage but were considerably heavier in weight and increased in size over their wild counterparts; and, finally, fully wild subspecies of *Anas platyrhynchos* collected from the marshes of Manitoba. It is difficult to say to what extent Gottlieb's results were influenced by the particular measures or tests that he utilized, but, in any case, his results do suggest that a higher proportion of an experimental population among Pekins can be induced to follow an artificial model than is the case among the wild mallards. When following does occur in the wild mallards, it appears to be of a higher intensity than in the Pekins. From these facts Gottlieb deduces that the spectrum of stimuli that can elicit following is narrower among the wild mallards than among their domestic counterparts. This conclusion was, indeed, supported by recent studies by Klopfer (as yet unpublished) in which the proportion of followers among the wild mallards rose far above that which obtained for the Pekin when a repetitive sound signal was added to the otherwise silent moving model of the duck. The Pekins apparently responded at a maximum level even in the absence of a sound, whereas the mallards require a simultaneous visual and auditory stimulation. Unfortunately, it is not yet known to what extent this tendency to follow a model is indeed a true measure of imprinting, if we consider imprinting to include a delayed sexual response. This is really a crucial part of the argument, since if the initial tendency to follow an inappropriate object does not lead to perverted sexual behavior, the reproductive consequences of the following response are nil. Certainly, following an inappropriate object may cause an increase in the loss by predation or other causes, and it is possible that this fact would tend for selection to operate against the following response quite irrespective of any sexual consequences of that response. To what extent barnyard mortality differs from that in the wild, however, is also quite unknown. It would perhaps be worth a bit of time to discover whether a barnyard strain of fowl given to following all kinds of moving objects suffers greater losses than are common among races that remain closer to their mother and are not easily induced to follow other animals or objects. In any event, we are obliged to conclude that domestication does affect the processes involved in the learning of an appropriate mate, possibly by changing the susceptibility to imprinting. The nature of the change remains unknown.

Evolution of Innate Behavior

It remains now to consider briefly some of the ways in which acquired constraints may become built into the organism. By this we mean how relatively plastic responses become a more predictable part of the species repertoire. Our argument will proceed by demonstrating that the acquisition of new behavioral traits occurring during learning involves physical changes in structure, probably within the central nervous system. We shall then argue that any structural change induced by the environment can also be induced, in the absence of the original environmental stimulus, by other events, including those termed genetic. We shall conclude by discussing the supposed causal relation between environmentally induced changes in the phenotype and those induced by genotypic mutation.

The search for the engram, that illusive trace of past experiences, has produced a plethora of hypotheses. Some of these investigations have assumed that physical changes in synaptic end bulbs account for the memory trace. Others have assumed the establishment of reverberating circuits that act to continually facilitate certain synapses along their paths. Reviews of these and similar hypotheses can be found in Milner (1960), Deutsch (1960), John (1967), and Horridge (1968). It would certainly be premature to claim that any of their hypotheses fully or convincingly explain how new responses are "built" into the nervous system. They do, however, reenforce our belief in a physical basis to memory that will ultimately be amenable to description. Thus the next question becomes, How do experientially induced alterations in the central nervous system's code become genetically fixed? That is, how can a response become "instinctive" in the sense of predictably appearing even when the organism is denied the company of conspecifics?

We can best approach this question obliquely by a more general consideration of the phenomenon called genetic assimilation by Waddington (1959). Suppose that the larvae of a certain species of fruit fly (*Drosophila*) are subjected to a mild heat shock. Of the adults that subsequently emerge from their pupae, a small proportion will be found to possess a minor abnormality in the pattern of venation found in the wings. If one selects these abnormal flies, breeds them, subjects their larvae to another heat shock, and then allows these larvae to pupate, the incidence of the abnormality will be found to have increased among the adults of this new generation. After a number of repetitions of this experiment through succeeding generations, it will be found that the condition appears with a high frequency even in the absence of any prior heat shock. Waddington explained this apparent genetic assimilation of an initially somatic response as follows: Selection was exercised for the genotypes giving a strong somatic response; subsequently, the genetic mechanism that assured a strong response was transformed into one whose threshold had been altered, providing for a response even in the absence of the environmental stimulus that was once required. The precise manner in which such a transformation occurs is never made clear. Consider, instead, the explanation of Stern (1958): Suppose a given trait (for example, abnormal vein patterns) appears regularly only in individuals who possess both recessive alleles at the genetic locus determining the trait. Although the heterozygote normally fails to show the trait, it may appear under a different set of environmental conditions (for example, higher temperature). If we take a small population of flies, say 10,000, and if the frequency of the recessive gene is low, say 10^{-6}, the probability is great that our population will contain no homozygous recessives; that is, all the flies will have normal wings. Heat shock, however, may reveal the presence of the somewhat more frequent heterozygotes to us, for the postulated trait is one that the heterozygote fails to show only under normal conditions. With the heterozygotes no longer indistinguishable from the dominant homozygote, it is now possible to establish a new population made up solely of heterozygotes. The frequency of the recessive locus or gene will thus have been increased and, ultimately, will reach such a level as to assure the presence of a relatively high proportion of recessive homozygotes. These, it will be recalled, will show the vein defect even under normal conditions, for example, no heat shock. Stern proposes this as a considerably simpler, and hence more satisfactory, explanation of the results of Waddington's experiment.

The point of greatest significance is that phenomena mimicking a Lamarckian form of inheritance (that is, an effect of somatic responses on the genotype) can be obtained without recourse to the unorthodox assumptions Lamarckian explanations require. In the case discussed above, the "new" trait was present from the start, though at very low and originally nondetectable frequencies. It is also possible for a particular phenotypic trait to appear de novo as the consequence of a gene mutation. If the mutant results in a phenotypic trait identical to that resulting from an environmental stimulus, the situation might also be interpreted, erroneously, as being due to Lamarckian causes. Thus, while heat shock might directly cause somatic changes that alter vein patterns, the identical patterns might also be the result of a spontaneous and causally unrelated mutation. Without a rather sophisticated analysis, an experimenter might never be aware of the lack of causal relations between the mutational event and the treatment he administered. This type of genetic mimicry has long been known as the Baldwin effect (Simpson, 1953) and doubtless explains many instances of alleged Lamarckianism. For example, assume that a given pattern of behavior is repeatedly learned by individuals of a particular species, the behavior pattern possessing an adaptive value, and that a selective advantage is conferred by a more rapid development of the behavior. A mutation that produces a change in the nervous system such that the previously learned pattern now appears prior to the situation in which learning usually occurs will then replace its normal allele. Thus evolutionary forces have led to the seeming transformation of a learned act into an "instinctive" act, though not through any mysterious or mystic forces. Indeed, it has been a case of simply substituting a specific genotypic adaptation for a more general genotypic adaptability.

The discussion of the Stern–Waddington controversy is relevant here because it illustrates the existence of more than one pathway to a particular goal. When we turn to the question of the specific relationship between any genome and an act, this multiplicity of paths becomes an even more crucial consideration.

Consider the fact that if chicks of a yellow and of a black variety are reared together and their individual preferences for one or the other variety are then determined, of chicks that do evince a preference, most will select chicks of their own hue as companions. These are animals that have reared together under identical conditions. Therefore we say that the differences between them are genetically determined (Kilham et al., 1968). It is important to note that this is not the same as saying that the preference for own kind is genetically determined. Why not?

Since we still know little about color preferences in domestic chicks consider instead the color preferences of the laughing gull (*Larus atricilla*) (Hailman, 1967). Color preferences in gulls depend largely on the presence and distribution of oil droplets in the retinal cells. These droplets act as band filters of varying widths and may excite or inhibit responses. How is the distribution and appearance of the droplets controlled by the genes? The most reasonable assumption is that at a particular locus of a chromosome a synthetic process is initiated that leads to the formation of enzymes or "messengers" that react with substances derived extrachromosomally to produce other substances that lead (eventually) to formation of certain oil-soluble pigments. Obviously, this description is but a small distance of the way to a color-coding and -perceiving mechanism. But the question "What in fact has been inherited" is clearly not susceptible to a simple answer.

The gene must be viewed not as a repository of data or a blueprint from which an organism can be constructed, that is, as an inchoate homunculus, but rather as an information-generating device that exploits the predictable and ordered nature of its environment (or, as Schroedinger, 1951, termed it, a device that feeds on negative entropy). This view accords well with current models of gene action, such as those advanced by Jacob and Monod (1961), among others, or that of Waddington (1966). A segment of the helix specifies a particular species of RNA that ultimately, and in an appropriate environment, leads to the synthesis of a particular enzyme, which, in turn, may repress or activate further synthetic activity by that portion of the helix or repress or activate another segment.

The stereotyped species-common feeding behavior of gull chicks (*Larus* sp.) affords another example of the impossibility of locating acts on genes. Typically, gull chicks, as previously mentioned, remain near the nest after hatching and are fed by the parents, which regurgitate semidigested food before them. The chicks peck at this in a typical fashion. Analysis of the pecking of the chicks shows that the movement consists of four discrete elements: a forward and upward movement of the head, an opening of the bill, a rotation of the head, and a push with the legs. There seems to be a considerable inter- and intraindividual variation both in the duration of these different components of the total movement as well as in their synchronization. Rearing the chicks under conditions of darkness and force-feeding them, thus denying them many of the usual visual and motor adjustments of normally reared chicks, affected some but not all of these components. Similarly, the chicks that were kept in the dark until an initial exposure to a variety of simple models that mimicked one or another of the characteristics of the parental head and bill were shown to respond differently to quite a number of features of the model. The experiments with models show that the important features of the parental head include the figure–ground contrast, the orientation (whether vertical or horizontal), the diameter of the simulated bill, its rate of movement, and its color;

many of these attributes could be manipulated in a compensatory manner without affecting the overall rate of pecking. The final outcome, the contact of the chick's bill with the bill tip of a parent or the food, may appear to represent a relatively simple and stereotyped behavioral response, but in fact it is a response that is composed of a multiplicity of movements and choices. Further, there is no reason to believe that in the forward movement of the chick's head the same muscle bundles are invariably involved—and, even within one bundle, different fibers doubtless fire at different times. Indeed, the more closely one approaches the molecular level, the more probabilistic and nondeterministic must our description become. The gull is truly a system that "examines" its own output (and adjusts it, too).

Behavior, however stereotyped, is the outcome of epigenetic processes that are expressed in the context of a stable environment. Actually, greater developmental stability is assured in this manner than by a biochemical homunculus. An epigenetic system is buffered and self-correcting at many points, as a homunculus cannot be. At the same time, the system is far more responsive to changes in environmental conditions than a homunculus would be. (Note, particularly, the reference to this labile aspect of epigenetic systems by Waddington (1966) and his discussion on canalization.)

A *gene* refers to inheritable differences. But, while it is nonsense to talk about the inheritance of behavior, it is true that behavior may be more or less stereotyped. One can imagine a continuum, with acts of perceptions or responses ranging from highly plastic and variable at one end to others that are highly constrained or stereotyped. Our inquiry is directed toward the following questions: How does behavior that falls on one end or the other of this spectrum develop? What was its evolutionary history? What are the mechanisms that underlie it? What functions does it serve? The answers to such queries reveal interesting differences between those kinds of behavior that we know to be highly plastic, those falling at one end of the continuum, and those kinds of behavior that we know to be more stable (less flexible), falling on the opposite end. Even the most stereotyped patterns of behavior upon which the barriers to hybridization depend will not be found in the chromosomes.

FIVE

How Are Communities Organized?

Groups of animals of the same or even of different species are often more or less loosely organized into social assemblies. These aggregates may be called communities. They vary in structure from the rigid castes of bees or termites to the seemingly casual flocks of tanagers (family Thraupidae) that an equatorial bird watcher cannot but espy. The nature of community structure and its origins and functions pose questions of interest.

The concept of community has been rather specifically defined by a number of ecologists, though the definition has changed somewhat from time to time and varies from one ecologist to the next. Odum (1959), for example, considers a biotic community as "any assemblage of populations living in a prescribed area or physical habitat. It is a loosely organized unit to the extent that it has characteristics additional to its individual and population components"

(P. 245). Among botanists there have generally been four different approaches to community study, some of which depend on physiognomy and others that depend on the dominant species or characteristic species or similar criteria. Whittaker (1956) has attacked the entire notion of communities, claiming that it is merely a convenient rationalization for the ecologist and not a true natural entity. This argument was based on his analysis of certain plant and plankton "communities" and his demonstration that the species characteristic for one area did not change concordantly as one proceeded to another area. The factors that limited distribution of one species of this so-called community differed from those that affected the distribution of another member species. The individual component species of this community altered their frequencies discordantly along environmental gradients. According to Whittaker, the conclusion that one must draw from his work is that the concept of the community as a discrete organized natural entity is largely false. Needless to say, this conclusion is disputed by other ecologists. In the present discussions, however, I am really concerned with a concept of community that differs radically from those of either the traditional ecologists or their antagonists. I am concerned with the local communities comprising groups of animals, the members of which react socially to one another. This is a much more limited and restricted sense of a community. Perhaps it is unwise to use the term to designate such a group, but if the meaning is made explicit, misunderstanding should be avoided. *Local community* may include many different species, but I exclude those whose interactions are solely of a prey–predator nature as well as those responsible for fatal parasitism. Nonfatal parasitism, such as that practiced by the cowbirds (*Molothrus* sp.) or the European cuckoos (*Cuculus canorus*), is included, and the reasons for the inclusion of this type of parasite should become clear as we focus on the specific problems involved in a study of community organization.

SOCIAL ORGANIZATION AND SOCIAL STATUS

It is in the monospecific communities lacking individual territories that the most rigid type of community organization is apparently found. It may be a linear dominance hierarchy, characteristic of a great many fowl. In this situation a single bird dominates all the others; another, subordinant to the first, dominates all those but the one bird above him, and so forth, down the hierarchy. The hierarchical order determines priority at the food and water dish and the selection of roosting sites or the choice of females for copulation. The entire spatial organization of the community follows from the dominance relationships. Among pigeons (*Columba* sp.), a rather similar structure exists, though here it is peck rights rather than absolute peck dominance that obtains, and the possibility of shifts in the hierarchy is greater. In these situations where dominance hierarchies determine social structure it is clear that individual recognition of conspecifics must take place. The dominance hierarchy is fairly stable, and the removal of a fowl from a flock and its reintroduction some days later will not always require this reintroduced individual to reassert its position. If it is a dominant bird, it may be recognized as such even after it has been gone for some days (Allee, 1931).

In some species, where the hierarchy is not so stable or so rigidly established as in the domestic fowl, the actual leadership of the group may rotate without respect to dominance relations. Allee et al. (1947) have found this to be the case for a flock of Pekin ducks (*Anas platyrhynchos*). It is also true for the young of several other species (Klopfer, 1959a). Among the young of those birds studied, no evidence of any sort of dominance relation could be obtained during the first two weeks after hatching. (W. Smith, 1957, has also noted this fact.) Yet groups of ducklings clearly followed "leaders." It was apparent that the leader is whichever duckling shows any kind of directed movement or "seems to know where he is going." This fact can be demonstrated by adding ducklings that have been trained to approach a particular object (a loudspeaker emitting a repetitive "komkom" chant) to groups of nonresponsive ducklings. The response of the group in the absence of the trained leader is first noted, the leader is added, and, finally, he is again removed. Repeating these tests with groups of various sizes and different leaders allows one to demonstrate that so long as there is one leader to every three ducklings the group will respond in unison, following their leader. Ambivalence and uncertainty develop when more than three ducklings are grouped with a single leader. This development was interpreted as an indication that with a larger group the probability had arisen that some duckling other than the leader would be first to show directed movement. The actual results are given in Table 5–1. Taken together, they would indeed appear to confirm the notion that among ducklings neither sex nor size nor color determines leadership, though as the animals mature these variables will clearly affect the degree of dominance.

In some monospecific communities, organization does not depend on a linear dominance hierarchy but rather on the existence of a polymorphism of the sort characterized by bees, ants, and termites. In these instances, castes exist that are physically different and that characteristically engage in different kinds of activities.

TABLE 5–1

Social Facilitation in Ducklings*

The fractions represent the proportion of groups of ducklings of varying composition that responded *uniformly* to a sound signal to which only the "leaders" had been trained to respond.

TOTAL SIZE OF GROUP	RESPONSES WITH 1 LEADER	RESPONSES WITH 2 LEADERS	RESPONSES WITH 3 LEADERS
2	4:4		
3	6:6		
4	5:6		
5	2:8		
6	0:2	5:5	
7	0:3	5:5	
8	0:1	3:5	
9	0:1	1:3	
10	0:1	2:3	
11		3:3	3:5
12		1:3	4:5
13		1:3	3:5
14			1:5
15			1:5
16			1:5

	FOLLOWER RATIO 1:3.3 OR MORE	FOLLOWER RATIO 1:4 OR LESS
1 leader	14:15	2:15
2 leaders	13:15	8:15
3 leaders	10:15	3:15
Total	37:45	13:45
95% confidence intervals	0.62 to 0.92	0.17 to 0.45

* After Klopfer, 1959b.

Much is known about the genetic and hormonal basis for the polymorphism, but the opportunities this situation offers to students of behavior have been inadequately exploited. Clearly a queen bee and a worker bee exhibit different patterns of behavior and different responses to particular stimuli. This would afford an excellent opportunity for gaining more insight into the nature of the neural substrates of behavior. Although much attention has been focused on endocrinological and nutritional differences between queen and worker, differences in the structure and function of portions of the central nervous system have not yet been examined. It is possible, of course, that the different behavior patterns of queens and workers can be explained simply in terms of changes in thresholds, thus allowing a fundamental similarity of the structures in the central nervous system. While this cannot be precluded as a real possibility, at the same time one cannot take an identity in neural function and structure for granted.

Polymorphism imposes added problems in species and sex recognition quite beyond those already considered in the foregoing discussion of sexual dimorphism. Where the polymorphic forms remain in contact with one another and freely interbreed, it is clear that their perceptual mechanisms must be attuned to recognize or respond to a wider variety of stimuli. A honey bee (*Apis mellifera*), for example, must be able to make finer discriminations than "conspecific" or "alien." One wonders whether this necessity has otherwise restricted the plasticity or range of responses of which bees might be capable. For example, suppose we make a comparison between related species of insects, some of which evidence polymorphism and some of which do not. This might allow one to determine whether the additional burdens on the processes of species recognition that are imposed by polymorphism have somehow lessened the species' range of responses to other elements of their world.

Polymorphism—perhaps *polyethism* is more appropriate—need not depend solely on the genetic and structural differences displayed by insects. African hunting dogs show a division of labor (and presumably status), too, some hunting and some guarding the young. Unlike the insects, the roles of hunter and baby-sitter are interchangeable. Here the community's structure is much more difficult for the observer to perceive. Roles are neither fixed nor morphologically determined. The questions raised below concerning role identification are especially crucial here.

Finally, community structure may depend neither on dominance relations nor role differentiation but on territorial prerogatives. Either individually or in concert with others, an animal may have more or less exclusive use of a particular area, for either some one or several purposes or for a longer or shorter fraction of its life. Territoriality does not eliminate social interactions but limits them (mostly) to the periphery of the territory or some distance from them. Hence the need for roles or dominance is reduced; in his castle, each man is a master.

An incredible amount of nonsense has been written about territorial behavior, largely because of the assumption that a robin (*Erithacus rubecula*) defending his territory has emotions (or something fundamental) in common with a human farmer threatening a trespasser. In fact, there seem to be as many kinds of territories among animals as there are species; they vary with respect to being solitary or communal and in size, temporal stability, and function, to name but a few of the major parameters. Among two closely allied species, one may be territorial and another not. The question of whether there is some ground common to all territories is referred to in Chapter Six (also note Klopfer, 1969a). The relevant point here is that organization in space represents one other form of community structure.

Communal territories may provide for an overlapping of categories. Prairie dogs (*Cynomys* sp.) inhabit territories. The territories are subdivided into coterie territories, which are held by small family groups and display an intrafamilial hierarchy (King, 1955). The gross features of community organization are fixed and maintained through the clear-cut territorialism shown by the family units. In the case of the prairie dogs the stability of these territorial boundaries is pronounced. Unlike the situation in many other animal communities, as density rises above tolerable levels it is not the young animals that emigrate but the old ones. Thus the fixity of the territorial boundary implies that there is a training of the young by their parents, a training that involves teaching the offspring the extent of the family territory. When tradition or a mode of behavior is imparted to the young by the adults there is a transfer of information, which is what we mean when we talk about communication. How does communication in these nonverbal animals take place? How generally does communication in any kind of nonverbal animal take place (for that it does take place we can be certain)? We shall have to return to this problem.

THE ATTAINMENT AND RECOGNITION OF SOCIAL STATUS

How does one determine his place in a dominance hierarchy? Schoolboys may fight it out, but this is not so often true among animals. Once a hierarchy has been established, the ability to recognize individuals greatly reduces the number of conflicts. This factor poses problems that are largely unsolved. Natural processes will favor the evolution of intraspecific and individual recognition mechanisms, for thereby the number of conflicts and thus the amount of time and energy wasted can be greatly reduced. That individual recognition does take place in a great many different kinds of birds, to say nothing of mammals, cannot be disputed. The literature is replete with examples, such as the phenomenon of penguins (family Spheniscidae) returning to the creche (where the young of many hundreds of parents band together) and feeding their own young, selecting them from a mass of milling youngsters. Fur seals (*Callorhinus* sp.) clamber over hundreds of individuals in the rookery in their search for their own offspring. Adult gulls (*Larus* sp.) in a colony bring food to their own young and not to another's.

In some monkeys relative status may be permanently fixed within the first hour of meeting (Bernstein and Mason, 1963), presumably by the perception of physical attributes. That such perception does occur has been clearly established: Young rhesus monkeys (*Macaca rhesus*) select social peers as play-partners (Pratt and Sackett, 1967); even isolated rhesus monkeys respond differently to photographs of threatening and nonthreatening monkeys (Sackett, 1966). In animals other than primates, individual discriminations of rank by scent are known in fish (Todd et al., 1967) and rodents (Kalkowski, 1968) and by voice in birds (Lemon, 1967; Emlen, 1971). In monkeys the communication of affect in more subtle ways is even possible across a television screen displaying no more than the face (Miller et al., 1963, 1966). Kneutgen (1970) records the amusing example of a bird whose illness led it to adopt a perverted posture that others interpreted as a sign of dominance. The sickly one was thus deferred to as the alpha animal.

Nevertheless, if we know that the perception of physical attributes leads to recognition of status, we know little enough about which attributes and their relative importance. Not all animals focus merely on the angle of a tail.

The attainment of a particular rank may be influenced by one's forebears. In free-living rhesus monkeys, the sons of high-ranking females remain in the central portion of the troop and generally attain higher rank than the sons of lower-ranking mothers, who move to the periphery (Koford, 1963). Marsden (1968), too, found that the offspring's rank usually followed its mother's. Rank may also be related to the level of sexual activity, which in turn depends on hormone levels. Injections of testosterone can raise the position of a subservient male in the hierarchy, just as estrogen may lower the position of a dominant male. However, the level of the sex drive itself, at least if measured in terms of purely sexual activity, cannot alone explain the position of an individual in the dominance hierarchy. It is possible to show that the most dominant male in a moderately sized flock of fowl will not necessarily fertilize the majority of the hens. Indeed, by the very fact of his dominance he may have to spend so much of his time reasserting his position that he leaves the field wide open to subordinate males. Certainly in other animals such as the fur seals (*Callorhinus* sp.), where a dominance hierarchy exists, there is as yet no evidence to suggest that the position in the hierarchy is invariably correlated with the level of fertility or sexual activity. And in the wolf colony (*Canis lupus*) of Ginsberg et al. (1967) the dominant male clearly inseminated fewer females than some subordinates.

The foregoing instances afford examples of individual recognition. Presumably, the same process is involved in the recognition of individuals standing higher or lower in a dominance hierarchy, but we know all too little about what the cues are (whether visual, auditory, or olfactory) and how they are integrated. It is a curious fact, for example, that a gull (*Larus* sp.) can presumably learn to recognize its own young in a relatively short period of time, and yet, when imprinting takes place among the same species, we assume that what is learned are not the specific features of a particular individual but the class characters of a group of individuals. Why should there be this difference in the specificity of the learning when the situations pose similar problems, the problems of learning to recognize an appropriate object within a rather short space of time and in the absence of overt reenforcement? What is it that assures the highly selective nature of the learning that is entailed in the recognition of one's own young or of conspecifics above and below one in the dominance hierarchy?

While an experimental formulation of these problems awaits the future, it is possible to advance a few generalizations. First, it seems reasonable to suppose that learning processes akin to imprinting will be of relatively greater importance in individual recognition among precocial animals than among those whose immature physical condition assures continued contact with their own kind. Imprinting, despite its manifest advantages with respect to the speed of its completion, also poses dangers for an animal (and a species) in that it can easily lead to inappropriate fixations. Where a period of physical helplessness intervenes between birth and dispersal, natural selection can be expected to favor the slower-acting but less risky processes of associational learning. Second, as suggested earlier, the complexity of the relevant stimulus field will be a function of the organization of the perceptual apparatus. Insects are presumably more limited to the perception of successive stimuli than mammals, whose nervous systems allow more readily for the simultaneous analysis of stimuli in several modalities. Thus specific releasors or sign-stimuli would be more likely to offer the relevant cues in the learning of conspecifics among insects than they would among mammals. The latter, presumably, focus more on Gestalt characters, features characteristic of the entire organism. Finally, where much sexual dimorphism exists, as in many ducks (*Anas* sp.), the effects of early learning processes should be less stable than in monomorphic forms, as the geese. In the latter, imprinting of the young to the mother can be relied on to provide the young with all the information required for the recognition of a conspecific mate later in life. Consider, however, a species such as the wood duck (*Aix sponsa*) in which the male is very strikingly colored —distinct not only from the female in his plumage, but from all other species that occur in the same region as well. Imprinting of young to mother will hardly provide much of a basis for later recognition of males by females, and, indeed, it does appear that female wood ducks will breed with males of a variety of other species (Dilger and Johnsgard, 1959). Thus, in fact, whatever imprinting does occur, it does not fix the sexual responses of the female to the male of her species. The males, on the other hand, not unreasonably fix their attention on females of their own kind. These females, after all, do resemble their original mother or imprinting surrogate. This has led

Hailman (1959) to suggest that the striking color patterns of the male represent a device to promote learning on the female's part of the characteristics of the conspecific mate. The more distinctive the male, the more readily he can come to be distinguished from males of alien species. Thus we can conclude that, in all probability, it is in monomorphic forms that imprinting will be of primary importance in species recognition. An interesting test of this hypothesis could readily be made by comparing the sexual effects of imprinting in mallard ducks (*Anas platyrhynchos*) some of which are monomorphic (for example, Mexican duck, *Anas platyrhynchos diayi,* or Laysan duck, *Anas platyrhynchos laysensis*) and others of which are not. A corollary of this hypothesis is that in monomorphic species that are altricial, that is, go through an extended period of dependence on the parents, pairs, once formed, will be relatively stable (cf. Hamilton, 1961). Thus it is a well-known fact that the birds of tropical regions form breeding pairs more slowly than those of more northern regions and that, once formed, these pairs remain together for a longer period. They are also markedly less inclined toward sexual dimorphism (Hamilton, 1961).

In summary, we suggest that (1) as the degree of precocity of the young increases, so does the likelihood of imprinting-like processes determining species recognition; (2) with an increase in the organizational complexity of the nervous system, the role of Gestalt characters becomes relatively more important than that of sign stimuli; and (3) other things being equal, sexually dimorphic forms will depend more on processes other than imprinting for species recognition than will monomorphic forms. Ducks should show more imprinting than thrushes, but among both ducks and thrushes the importance of imprinting for assuring species recognition should be correlated with a lack of sexual dimorphism.

How does one explain, however, the ability of nest parasites, among birds, to recognize their own kind? *Aythya americana,* the redhead duck, frequently deposits its eggs in the nests of other species of ducks (Weller, 1959), as do the European cuckoos (several genera) and North American cowbirds (*Molothrus ater*). The last may form large species flocks after fledging, apparently discriminating their own kind from their foster parents and related species. One can surmise that this behavior has become possible only as the consequence of the evolutionary development of a perceptual mechanism so constrained as to react only to a specific configuration of stimuli. Such a device would roughly correspond to the innate releasing mechanism (the IRM) of the ethologists (cf. Thorpe, 1956). Some neurophysiological evidence for the existence of the type of stimulus-filtering process this IRM requires has been reported by Maturana et al. (1960). By recording the output of fibers from the retina of the frog's eye, they were able to demonstrate the existence of five functional classes of retinal elements. Of these five, three were responsive to movements and two to changes in illumination. The three movement-responsive groups were further differentiated into classes of elements that attended to the detection of convex edges, sustained edge detection, and changing contrast. Thus the eye of the frog is designed to allow discriminations between fairly complex visual patterns. It is, therefore, not too difficult to believe that in the cowbird (*Molothrus* sp.) a similar filtering or analyzing mechanism exists, one specifically attuned to the characteristics of the species, which, once stimulated, releases the appropriate flocking or breeding response. The demonstration of the actual mechanism, of course, awaits an ambitious experimenter. Similarly the evolutionary cause of some avian species relying on such mechanisms to assure species recognition while other closely related species apparently are provided with more plastic mechanisms remains an enigma.

SOCIALITY AND COMMUNICATION

Sociality apparently depends on information transfer or communication. Communication between organisms can occur through a variety of channels. Those that are most predominantly used are the visual, the auditory, and the olfactory senses. To one extent or another, however, the questions that we shall raise relate to all these modalities, so we need make no distinctions among them. We do need to distinguish, however, between communication on inter- or intraspecific levels, for there are differences in the communicative processes and the problems existing at these two levels. In the present context it is only intraspecific communication with which we shall be concerned.

To begin, let us examine some of the elementary facts about animal communication with which we are conversant. We know, first, that a great many animals produce signals that are understood by their fellows in the sense that the latter react to these signals in a predictable and biologically appropriate fashion. One alarm call of a mother duck (*Anas* sp.) produces a quick aggregation of her young and movement toward her; another leads to the immediate dispersion of the clutch. One call from a mother quail (*Colinus* sp.) will cause the young to crouch motionless beside the bush; a different call will cause them to come running toward her. We do not, by any means, know what the meanings are of all the various calls characteristic of some species. The wild turkey (*Meleagris gallopavo*), for instance, is one of the most highly vocal of the gallinaceous birds, and it seems reasonable to assume that the different calls it is capable of producing each have a particular meaning. The nature of these meanings has, as yet, not been uncovered. One of our first problems, then, has to do with the meanings of particular signals, and the second, and more important, has to do with the processes through which meanings come to evolve. These are problems not unlike those that face human linguists. It is remarkable that so few zoologists have focused on them (but cf. Marler, 1961). Admittedly, a good start has been made. Morris (1956), for example, in considering the signal function of feather postures in finches (family Fringillidae), has pointed out how these postures can be interpreted as ritualizations of comfort movements. According to Morris, autonomic responses, including piloerection, accompany somatic responses. The former can then give rise to social signals. Besides the origin of social signals from autonomic responses, the nature of the displacement activities that occur when two conflicting drives are simultaneously aroused may be determined by autonomic responses, depending on the pattern of activation at the moment the conflict situation arises. The displacements may then secondarily evolve a signal function.

A study of the evolution of the complex signaling system common to some insects has been contributed by Blest (1960), who has devoted time to an analysis of the evolution and ontogeny of the settling movements of a number of saturniid moths (family Saturniidae). In his New World saturniids, he found that the strength of the rocking response performed by the moths upon settling was influenced by the duration of their previous flight. The effect of the previous flight was unaltered even by such drastic procedures as prefusing the body with glucose or Ringer's solution. This factor belies a registration within the central nervous system of flight duration and a simple afterdischarge that is quantitatively related to the flight duration (at least insofar as the registration depends on metabolically induced changes in fluid composition). Blest also noted that flight normally elicits sexual responses and that these in turn inhibit settling. He therefore deems it significant that the postsettling dances that have been elaborated into a communicative function have evolved only in the functionally sexless castes among colonial insects.

The next step in tracing the evolution of such involved signal systems was taken by Dethier (1957) in the study of the honey bee (*Apis* sp.). He showed that a number of noncolonial insects will perform rudimentary dances upon feeding. In some flies the stimulus to dance is the taste of food. The intensity of the dance and the effectiveness of the taste stimulus are modified by the individual's nutritional state. Bees, of course, can communicate more than the fact that food has been found. The work of von Frisch (1950) and his students has confirmed that bees can convey information about the direction, distance, nature, and quantity of food. Thus many links in the chain from rocking movements of moths to the round and waggle dances of the bees remain to be forged.

Repeatedly confirmed in recent years, particularly by the works of Thorpe (1958), Thorpe and Lade (1961), Marler (1961), and Lanyon (1960), is the observation that there are local dialects among different populations of a given species. This factor results in a considerable variation in the quality of the song and sound signals of different individuals of the same species. The origin of such local dialects is not difficult to explain—no more difficult, in any case, than the origin of any regional variation in morphological characters. Vocalizations will ever tend toward those sounds or sound combinations that are most easily and efficiently uttered. The work of Zipf (1949) provides ample confirmation for this, at least insofar as human language and sounds are concerned. Since a great many components of the individual's anatomy go into sound production and since there is inevitably much genetic variation in the shape, size, and position of these components, it is to be expected that regional differences will arise in the structure of the sound-producing apparatus because of either adaptations to local conditions or genetic drift. These anatomical differences in sound-producing mechanisms will lead to differences in the kinds and quality of sound produced. Indeed, this concept, originally proposed by Waddington (1960), has been elaborated by Broshanan (1961) to explain the existence of many varieties of human languages and the differences between them. The same argument can be applied to signals in visual and olfactory modalities as well. To what extent the meaning of regional dialects can be understood by members of one or another local population is a problem that can easily be solved. We shall consider its implications for interspecific communication in the paragraphs that will follow shortly.

Of more immediate interest is the problem of how meanings are learned by the young of a particular species, as well as how the particular sounds themselves may be learned. In connection with the last point, unfortunately, too few species have been studied to give a complete picture. What has been noted is that there are considerable differences from one species to the next in the processes whereby the species' characteristic songs are learned. In common roller canaries (*Serinus canarius*), for instance, and in the European whitethroat (*Sylvia communis*) and the European blackbird (*Turdus merula*), the normal song is produced even when these birds are reared in auditory isolation from one another. Indeed, the blackbirds may even be deafened operatively and yet produce a song that is characteristic of their species. On the other hand, in the chaffinch (*Fringilla coelebs*) a predisposition to learn the correct song apparently exists only during the bird's first fall and spring. Thereafter, if the bird is isolated from its conspecific it shows no particular preference for any given series of sounds, although the gross pattern of the song that it produces will remain characteristically chaffinch-like. Also in this species, and this is unlike the canary, androgenic hormones will not enable a female to produce a male song. However the male learns to sing, his song is a trait not shared by the female, while in canaries, as we have indicated, this is not the case. Thorpe (1958) suggests that certain components of the chaffinches' songs are acquired by a means of a rapid form of learning similar to imprinting-type processes, whereas other components are acquired by more gradual learning processes. Some features, such as the duration of the song, are apparently characteristic of the animal whatever treatment it is accorded. Simms (1955) has claimed to have recorded a series of calls and responses between a brooding shorebird and young still in the egg, suggesting that external influences may be operative at a very early stage. Lorenz (1941) has indicated that much the same thing may occur in the case of some ducks, though neither Hess (1959) nor Klopfer (1959b) has yet been able to demonstrate that recordings of sounds played to the young while still in the egg will produce any kind of change in the bird's responses to these sounds. Among surface nesting waterfowl, apparently, a particular sound is learned only when it is linked with some visual stimulus so that it can in time substitute for the visual stimulus and elicit similar kind of response. Among hole-nesting ducks, at least those species thus far examined, this is not the case. Here a particular response can be linked to a sound signal in the absence of visual stimulation, a fact that is of no small adaptive significance. This adaptation occurs in hole-nesting species, such as the wood duck, since the first response of the young to their mother may be made at a time when the mother cannot be seen, the nest of these animals being recessed inside hollow trees. Thus a great variety of mechanisms apparently exists to assure that a bird will learn the calls characteristic of its species, and, aside from the weak generalization apropos the waterfowl, it is not possible to establish any general rules that specify when one type of learning can be expected to be found and when another. Even less can be said concerning the learning of the meaning of the sound apart from the learning of its physical characteristics.

If we return to the former point, how meanings are learned by the young of a particular species, we shall find that out speculations are scarcely at all inhibited by facts. In general, one can postulate that built-in constraints, the IRMs of an earlier section, may "define" the meaning of particular signals. Alternatively, particular classes of signals may lead to generalized or diffuse responses, perhaps merely an emotional arousal, with differentiation among members of the class and the development of specific responses occurring secondarily. Thus the young of many species of surface-nesting ducks will aggregate at the source of any of a wide range of rhythmically repeated sounds (Klopfer, 1959b). Should the sound source, if it is of an appropriate size, begin to move, the ducklings will follow it. Thereupon a generalized tendency to approach any rhythmic sound will become transmuted into a tendency to approach only a particular sound and, what is more, to follow the source of that sound should it move. The specific meaning "come to the source" is, in effect, attached by the duckling as the consequence of a visually induced following response. The moving object the duckling follows can be regarded as the unconditional stimulus; the act of following, the reenforcement (cf. the previous section on imprinting); and the auditory signal, the conditional stimulus. The last can then take on the "meaning" of the unconditional response. (Presumably just the reverse occurs in at least some hole-nesting species of ducks, where it is the sound signal that serves as the unconditional stimulus and the visual model as the conditional one.) Finally, and obviously, many meanings are imparted by a direct and repeated conditioning or training of the young by their parents. This form of learning is, of course, contingent on the existence of an extended familial relationship and would not be expected to be especially widespread throughout the animal kingdom. It can be presumed to be of particular importance among mammals whose social groups are relatively stable. One gets the impression, for example, that the young of many Canidae (dogs, wolves) learn the meaning of specific facial expressions or vocalizations only gradually, repeated bites being necessary before the pup discovers that not every form of the growl is an invitation to play.

In local communities consisting of different species the several different problems of communication become inordinately complex. Not only is it necessary that the individuals in such a community continue to learn to recognize the signals appropriate to their own species, but also they must now be able to distinguish these from the signals of the other species and to make the appropriate responses to these interspecific signals as well. Some of these problems have been considered in detail by Marler (1961). For example, sounds that facilitate locating their originator in space, which are important in breeding display and care of the young, tend to have physical characteristics that are closely adapted to their actual function. They are segmented or repetitive, their duration is brief, and they are composed of relatively low frequencies with much of the energy coming at the beginning of the sound. These are characteristics that greatly aid localization of the sound source. At the same time, such calls can be highly varied and thus can maintain a great deal of interspecific difference. Other sounds commonly used are intended to give warning and at the same time to confuse the hearers about the location of the sound source. In mixed flocks, for example, if a predator threatening any one of the component species is sighted, the warning call given by the lookout must be recognizable by all members of the community if it is to serve an interspecific social function. At the same time it must have characteristics that do not particularly endanger the signaler. Marler points out that signals with these characteristics are generally pure tones in the region of six to eight kilocycles per second, calls with no discrete beginning or end and with a fairly equal distribution of the energy. Sound spectographs (Marler, 1961) have shown how similar such warning calls are among widely distributed animals, from birds through mice.

To what extent interspecific recognition of other calls (besides warning calls) is possible poses a problem that well merits some attention. Frings et al. (1958) noticed that minority groups that may flock with other species are more likely to learn the appropriate reactions to the signals of the majority species than conversely. How generally this is true for other organisms is not known. Frings et al. (1958) showed that herring gulls (*Larus argentatus*) and eastern crows (*Corvus brachyrhyncos*) would disperse or aggregate in response to recordings of the appropriate assembly and flight songs of their species. He then showed that, upon exchanging sound recordings with his European collaborators, the distress calls of French corvids when tested on eastern crows in Maine in the summertime provoked no response in these animals. Interestingly enough, when the same calls were tested in Pennsylvania they produced an approach response. A test in Pennsylvania in the winter resulted in the same negative results as had previously been obtained in Maine. The reciprocal experiment with eastern crow calls being played to French crows demonstrated that the French crows reacted not at all to the flight dispersal call of the eastern crow, though the assembly call did provoke an initial approach followed by flight. Apparently, the French crows are less provincial than at least some eastern crows. Later Frings proposed an explanation for the difference in the responses of the winter and summer populations of the eastern crows to the French crow calls. The crows in Maine during the summer migrate to Pennsylvania during the winter. The Pennsylvania crows, in the meantime, migrate much farther south. Thus in summertime the crows that are in Pennsylvania are really animals that have had a winter experience in regions farther south where they have had contact with other species and subspecies of crows. The Maine crows, however, who never get any farther south than Pennsylvania, are regularly isolated from other populations. Frings proposed that the experience of the more southerly populations of crows with very different dialects has reduced their provincialism, which reduction, in turn, accounts for their increased tendency to respond, even if not entirely appropriately, to the calls of their congenerics in France.

This reaction does, however, raise a very intriguing problem as to whether the spectrum of signals to which an animal can learn to respond can be broadened as a result of more experiences with those signals during particular portions of the animal's life. If we may reason from human analogy, this hypothesis is credible. The ability to learn foreign languages is generally greater in children who experience a great variety of languages at an early age than in the case of children denied this experience. Indeed, Penfield and Roberts (1959) have argued that shortly before puberty some substantial changes take place in the neural organization of the speech centers that render incomparably more difficult the complex task of assimilating a language and associating meanings with sounds and the motor patterns required to produce them. It is not unreasonable to imagine that much the same situation exists among other animals, and the experiment whereby animals are reared under conditions affording them different degrees of social stimulation followed by a test of their ability to learn social signals of other species is one that is crying out for a performance.

Where two or more related species occupy the same or adjacent regions, there will be even more intense selection pressure favoring the ability to distinguish between inter- and intraspecific signals. This hypothesis follows from the assumption that hybrids between two distinct species will be at a selective disadvantage relative to their parents. Hence, natural selection will lead to the evolution of barriers to hybridization, among which will be a keener discriminatory ability. Alternatively, the signals produced by allied species may become more distinct in the area of overlap. That discriminatory ability may increase among species exposed to a greater variety of signals is suggested by the results of Frings et al. (1958). That the signals of sympatric species do indeed show their maximum divergence in their characteristics in the zone of overlap is even more clearly known from, inter alia, the work of Marler (1961) with birds and Blair (1958) with amphibia. This phenomenon, known as character displacement, has already been discussed in Chapter Three in a somewhat different context. (For a more general review of communication, see Sebeok, 1968.)

Sound signals are not the only kind of signals exchanged in mixed species populations. The visually perceived signals, or olfactorily perceived signals, may play as great a role, though it is clear that these may be much more difficult for us to study. As one might expect, the ability to read the signals of an alien species is at least partly a function of the degree to which the animal in question normally associates with animals other than those of its own kind. It is clear that an inexperienced puppy confronted with a cat with back arched and mouth twisted into a hissing snarl is not going to be able to interpret the signals from that cat in the way in which either an experienced dog or a cat-loving human will. Truly enough, it does not take long for a puppy to learn what this particular set of visual stimuli means. But how many different sets of such signals can such a puppy learn if it is reared with other animals? Will it learn to respond appropriately to the warning cough or bark of a species of deer with which it has been living in association? The experiences of Thompson and Heron (1954) have some relevance to this question for they demonstrate that the variety of stimuli impinging on a young puppy during various stages in its development influence, to a very marked extent, the emotional ability and curiosity of that animal when it reaches maturity. Puppies reared under conditions of isolation in a rather homogeneous environment proved far more frightened of strange objects to which they were exposed later and far less likely to explore a new environment into which they were placed than were puppies who were reared in a lab or in a home. Whether or not one can extrapolate such results to mean that a similar situation obtains with respect to the ability to learn the sign signals of alien species cannot be answered, except through the performance of the appropriate experiments.

THE FUNCTIONS OF SOCIAL ORGANIZATION

A number of putative functions of communities have already been alluded to. We cannot, however, ignore the suggestion by Wynne-Edwards (1962) that the primary purpose of social organization is in preventing overpopulation and irrevocable destruction of vital resources. Wynne-Edwards postulates the following: Food ultimately limits the number of animals that can coexist; a contest for food must ultimately lead to overexploitation and destruction of the habitat or starvation (note Hardin, 1960). This is avoided by the substitution of "conventional rewards," symbols of food such as social status or priority of place that are scarcer than food. The competition for these conventional rewards acts to maintain the population at an optimum level, below that level at which food resources would be limiting.

The difficulty with this notion is that it defies the action of selection, which always pushes organisms along the path of maximum fitness. This is measured by the relative proportion of individuals in a future population that can be attributed to their progenitors. Thus, if 49 percent of the individuals alive in 1980 can trace their descent to A and 51 percent to B, the latter is the more fit. If there are no changes in relative fitness, B's descendants will eventually dominate the population. Wynne-Edwards surmounts this difficulty by invoking the idea of group selection. There is an obvious benefit to a group to restricting population growth, so Wynne-Edwards believes this advantage can override the individual disadvantage of restricting reproduction.

In his "The Tragedy of the Commons," Hardin (1968) succinctly portrays the human analogue to this scheme. The "commons" are the communal pastures of old England. Suppose that a common could just maintain 100 sheep, equally divided among ten farmers. If one farmer adds one additional sheep, the pasture will begin a slide to inevitable destruction. The cost of this will be equally borne by all ten farmers. However, the temporary extra profit from the extra sheep will accrue solely to the benefit of the one cheater. Hence the tragedy: Cheaters do prosper.

Natural selection can operate only on reproductive units, not on portions or aggregates of them. Hence the notion of group selection as envisaged by Wynne-Edwards remains a vague and mystical scheme (see Wiens, 1966, for a detailed critique).

None of this denies the existence of a form of group selection dependent on kinship. Since siblings must each possess the same number of parental "genes," the fitness of a parent is uninfluenced by which of several siblings survives. Two offspring possess as much of one parent's genotype as does that parent, and three possess 50 percent again as much. Hence a single parent's fitness would increase if his death provided for the survival of three offspring, rather than the converse. The evolution of individually harmful behavior that is beneficial to a group ("altruism") can thus be explained (Hamilton, 1964). Specifically, if the gain to a relative of degree r is K times the disadvantage to the altruist, selection will favor the altruistic "genes" if K is greater than $1/r$. This can explain many of the phenomena that group selection purports to deal with. More to the point, the possibility of evolving altruistic behavior is initially dependent on at least a rudimentary social organization. Perhaps this should be seen as sociality's major function.

Social organization provides more immediate advantages than this enhanced evolutionary potential. It can increase breeding success by providing for mutual stimulation and, by synchronizing reproduction, for protection from predators (Darlington, 1959). It obviously sets the stage for observational learning, which provides for traditions and the behavioral precursors to culture. The occurrence of observational learning, alluded to earlier, and its dependence on stable social relations, is evident. For example, Chesler (1969) found that kittens learn much more readily if the "teacher" was their own mother rather than some other cat.

Perhaps one of the most significant aspects to sociality, and its least valued, is its promotion of diversity. The advantages of niche diversification among competing species has been extolled earlier. Much the same argument can be advanced for intraspecific role diversification—which may or may not be associated with genetic polymorphism. The prevalence of schizophrenia in Western societies has been seen as evidence of a genetic polymorphism that continues to exist because of the advantages conferred on certain of the morphs (Huxley et al., 1964). At a nongenetic level, the sexually defined differences in the roles of Western men and women may have allowed for the preservation of unpopular viewpoints: At least until the advent of women's liberation (and this comment is not intended to be pejorative), women were permitted to espouse causes forbidden to men (Margaret Mead, personal communication). It is probably not coincidental that major social reforms in the United States have so often been initiated by women. At the nonhuman level, R. E. Eaton (personal communication) has reported that some lion mothers are very "good"—they stay close to their cubs. As a consequence, they are not very good hunters. The "bad" mothers leave their cubs but bring home food. By dint of their sociality, good and bad mothers together do better than either separately.

The appearance of social behavior provides for positive feedback. First, it allows for polymorphism, role differentiation, as noted above. It also allows for observational learning and the development of traditions. Traditional activity, in turn, allows for modification of the environment by the construction of durable nests or the invention of clothing, which, in turn, alters the selective advantages of various genes (Caspari, 1963). This theme has been elaborated in various ways by Garn, 1963, and Crook, 1970. We shall return to it in Chapter Six.

SIX Ecos and Ethos

Do studies of the ecology and behavior of animals throw light on the origins and significance of human behavior? We have not hesitated to extrapolate from mouse to man when dealing with problems of morphology. Phylogenetic reconstructions, the application of the principles of homology and analogy, are the mainstays of classic biology. The application of a similar approach to human behavior has been scarcely less popular, whether by biologically oriented psychoanalysts (Freud) or applause-conscious dramatists (Ardrey). What are the limits beyond which extrapolations from other organisms to man become unsound? Are there limits? Or does the ecological diversity of our planet preclude any grand generalizations? It has been cogently argued that the prevalence of convergences in behavior, the lack of structural correlates for behavior, and the multiplicity of the causal pathways underlying behavior prevent the meaningful application of the morphological concept of homology (Atz, 1970). Much the same point is implied in the studies of human social origins by Crook (1970), Struhsaker (1969), and Gartlan (1968), among others. Not only the impossibility of homologizing limits us: That most human of capacities, language, also may obfuscate and confuse. Let us consider three problems of interest to man that are relevant as well to other animals:

1. How might studies of animal sociality relate to studies of man? Specifically, we shall examine territorial and aggressive behavior.
2. How did mortality and man's conception of death arise?
3. How has man come to the concept of beauty?

In dealing with these specific topics, the limits of homology and language will become apparent, but hopefully so, too, will methods of surmounting these limits.

IS MAN "TERRITORIAL"?

In Chapter Two, it was suggested that territories are exceedingly diverse in nature and function. Even if the possessiveness certain men show toward real estate is termed *territoriality,* it is doubtful whether this behavior has any but the most superficial features in common with the territoriality of a bird or prairie dog (*Cynomys* sp.). Indeed, among closely related species of one genus of bird, one may be territorial and the other not (Brown, 1963). This argues against territorial behavior being the expression of some deep-seated structure, an incubus man cannot unload, and also suggests an answer to the question of origins. Brown (1964) proposes that we consider territorial behavior as site-dependent aggressiveness. Aggressive behavior itself is likely to occur and be maintained only where this enhances survival. The aggressive behavior could be linked with the defense of any or all resources. Whether it is employed depends on the dependability of these resources, that is, their availability and accessibility to each individual, and the cost (in time and energy) of obtaining and defending them.

> Too much aggression in the absence of a short supply of the disputed requisite would eventually be detrimental. Consequently, a balance must be achieved between the positive values of acquired food, mate, nesting area, protection of family, etc., and the negative values of loss of time, energy,

> and opportunities, and risk of injury. Where this balance may lie in any particular species is influenced by a great variety of factors. . . .
>
> Within the population those individuals with the optimal balance of the genetic factors working for and against a particular form of aggressiveness. . .would [become] the norms of the population. [p. 162, Brown, 1964.]

If a territory does evolve, its nature depends on the economic considerations. Marine birds feed from the seas; for these birds defense of only a small area to be used for mating and nesting may be worthwhile. This contrasts with terrestrial birds for whom the economic variables predict a different outcome.

But, what about man? What does the economic argument predict?

In general, it appears that mobile animals, whose resources are widely scattered or else low in energy and abundant (demanding much time for sufficient calories to be collected), cannot afford feeding territories. This is true of marine birds; it is also true of grazing mammals. Nor do the latter even require a nest site. Indeed, the only vestige of territorial behavior many grazers show is during the rutting season when the males may compete for females. *Homo sapiens* is not seasonally reproductive, however, and his earliest feeding behavior apparently depended on gathering widely scattered low-energy foods (Birdsell, 1953). Territories were likely of no more use to him than to a goat or gazelle. Whatever the origins of defense of property in man, they are unlikely to lie in primordial and unalterable habits ordained by selection imposed on his ancestors.

EVOLUTIONARY ORIGINS OF MORTALITY*

> I grow old, I grow old
> I shall wear the bottoms of my trousers rolled.
> —T. S. ELIOT (1930)

The lament of Prufrock must surely strike responsive chords in all of us. Growing old, aging, is an irreversible function of time that

* Reprinted from the Duke University Council on Aging and Human Development, *Proceedings of Seminars 1965–1969,* with the permission of the council. Based on a talk given October 22, 1968.

is common to animate and inanimate nature alike. Rolling one's trousers, on the other hand, is not only a prerogative of man but of particular societies of men: Prufrock's distress is the consequence not so much of aging as of senescence—and particularly of his *awareness* of senescence.

By *senescence* we specifically refer to an increased probability of death with age. In the concluding paragraphs are summarized the explanation of Medawar (1957) of the origin of that particular type of mortality table that spells senescence. Yet a digression to consider the reasons for our awareness of senescence and the larger problem of aging could be instructive, for we are aware of senescence—we fear it, and we fight it. Such a digression should reveal the existence of cultures for whom time is a cyclic phenomenon, endlessly repeating, and for whom aging as an irreversible, unidirectional process does not exist. Indeed, aging then becomes an essential part of a cycle of birth–death–rebirth, a prerequisite to life itself. Watts (1966) has developed this notion in a metaphor that likens man and the other organisms on this planet to the many-headed hydra. Each head may seem to have both a certain degree of autonomy and a finite life span. The biologically acute observer, however, recognizes each to be part of an immortal longer-lived whole, not unlike the leaves of a sequoia. The notion of own, human separateness, one from the other, may be an illusion fostered by our imperfect senses or our underdeveloped brains. It is also possible that out notions are shaped by inadequate language. Indeed, the role of the latter in conditioning our world view and in shaping our concepts needs much more attention.

For instance, consider the small boy who was asked by his teacher, "What is half of eight?" His carefully considered response was, "Why, either zero or three, depending on which way you divide it." (He draws a line horizontally through the figure 8 yielding two 0s, and another vertically, making a 3 and a Ɛ.) His seemingly perceptive response reflects a confusion of a symbol with the object for which the symbol stands (Bateson, in press). This kind of confusion is not only very common, but because of the manner in which we consciously manipulate our environment and because of the power of our technology, it produces effects that are disastrous. I could state this problem in another way: Natural systems have a complexity that is an order of magnitude greater than that of our descriptive systems or languages, possibly even greater than the complexity of our nervous system and perceptual apparatus. The result is that incongruities arise. When these are amplified by the power of our technology, there is in consequence a destruction of natural cycles. The existence of smog in the Los Angeles basin or that body of inland sewage known as Lake Erie are examples of such disruptions.

Another instructive example is the cycle of nitrogen on land (ibid.). Plants represent a store of much of the terrestrial, organic, reduced nitrogen. When a plant dies, it decomposes, forming humus, and, through bacterial action, the nitrogen is released in the form of inorganic, oxidized molecules. These molecules, acted upon by nitrifying bacteria in the nodules of certain plants and in the presence of oxygen, which diffuses through the soil, are transformed into organic, reduced forms and are the substance of plant tissues. With the approach of agriculturally minded men, plants are cut, harvested, and removed. After a time, the amount of inorganic oxidized nitrogen available in the soil for further plant growth is lowered to the point where plant growth ceases; the soil is depleted. In preagricultural times, plant-hungry men simply packed up their gear and moved to richer farm lands. As his numbers grew and the amount of available land shrank, man resorted to the technique of supplementing the nitrogen compounds of the soil by adding nitrate fertilizers. This once again permitted plant growth and a continuation of the harvest cycle. But this procedure allows a continuing depletion of the humus. As this occurs, soil porosity diminishes, thus restricting the amount of oxygen that can diffuse to the roots and therefore decreasing the proportion of nitrates that may be taken up from the soil by the plant roots. The solution, of course, was to add more nitrates to the soil; of course, then more nitrates leach into Lake Erie and similar bodies, increasing their degree of pollution. [It has been estimated that the amount of nitrate that leaks from one square mile of farmland in a year is equivalent to the amount of sewage produced by 750 people over the same period of time (Commoner, 1969, 1970).] Too bad! But, what is more, as the amount of nitrate added to the soil increases, more accumulates in the plants in the form of inorganic nitrates. This is not an altogether desirable consequence since the nitrate, once the plant has been cut and cooked, is rapidly converted to nitrite, which because of its interference with the oxygen-carrying capacity of hemoglobin (leading to methemoglobulemia) may be lethal. For instance, spinach harvested in Europe may have up to 2100 milligrams of nitrate per kilogram. Since the U.S.-Public-Health-Service-recommended maximum for infants in 12 milligrams per day, a serving of roughly one-fifth of an ounce of this spinach constitutes the maximum permissible dose. This even ignores the fact that nitrate is absorbed from other sources as well, especially drinking water. In this country, we might also contrast the amount of nitrate in organically grown lettuce, which is about 0.1 percent, with that grown in fertilized fields, that is, fertilized with synthetic fertilizers, which is six times as great. The point to be made in citing this example is that the facts of the specific chemical reactions alluded to above are well known; some of them have been well known for a century or more. What has not been known, and is still not fully appreciated, is that these reactions do not occur independently of one another. They are part of cyclic, interactive systems. The cycle of nitrogen in the water interacts with the cycle of nitrogen in plants, which in turn interacts with the cycle of nitrogen within our bodies. These are systems that feed back not only on themselves but interact with other systems so that perturbations in one spread to the other, often with much amplification. Somehow we have not perceived the cybernetic nature of processes involved in nitrogen metabolism. It is going to be essential to look more closely at the reasons for our failure to perceive the integrated nature of these and other cycles, including the cycle of life and death.

Implicit in most of the statements we make in daily discourse is the belief that the phenomena we observe are polarized; that is, they are ordered into actors and actions, subjects and predicates, or nouns and verbs (Hardin, 1956; Whorf, 1956). This is reflected by our language: "Men control the environment," "the cat purrs," "the grass grows." Indeed, all the Indo-European languages require such ordering. One might well ask, What is the distinction between nouns and verbs? Is it a difference in substance or stability? Scarcely, since such insubstantial, unstable phenomena as "lightning" or "wave" are considered nouns, whereas "persistent" and "continue" are stable verbs. The inadequacy of this polarization becomes most apparent when the sky darkens and our ears are assailed by sounds and we say, "It thunders," "the lightning is flashing." What is the "it" that is thundering? What is the lightning that is flashing? These are tautological statements; they would be quite absurd, except that our language allows us no alternative. The effect on biology of this polarized way of viewing the universe has not been trivial. Indeed, the nineteenth-century view has been of the organism as a combustion engine whose action was metabolism. What is the engine, then? Why the protoplasm, of course. Thus has developed the view of the organism as a stable, reasonably static entity within which certain activities occur: "Protoplasm metabolizes" (Hardin, 1956). The inadequacy of this view, of course, is quite apparent to us today. We recognize an organism as a point in space where certain reactions are occurring. That is, organisms are open systems, every atom of which is in dynamic equilibrium with atoms of the "exterior" universe. This—more accurate—point of view was undoubtedly much delayed in acceptance because of the limitations of language. "Protoplasm metabolizes" is equivalent to "ether undulates," "lightning flashes," and "it thunders." It is well to note that some languages are free of the particular constraints that impose these polarized views on us, though, of course, they contain other constraints that may be equally limiting. For example, the language of the Hopi does not require polarization, but on the other hand it does not allow objectification of quantities either. That is, one can speak of ten men but not of ten days. Ten days are not a real aggregate but a metaphorical one, and statements of quantity in Hopi cannot deal with metaphors. Similarly, nouns in Hopi exist only for a specific drop of water, not for water in general. The important point, though, is that the Hopi language probably produces a different world view (Whorf, 1956), though it is not necessarily a world view with greater congruence than that which is allowed by our own language.

Anatol Holt (personal communication) has come to the same conclusion, pointing out that while nouns predominate in our grammar they probably have no existence outside of it. Phenomena that we regard as represented by static, stable endities in which certain processes occur are figments of our imagination. If every atom of our being is in exchange with the environment, so that those atoms that constitute "me" can in a very literal sense be said at another time to constitute "you," we are not static objects. Or, to restate an earlier phrase, we are a point in space at which some reactions are occurring more slowly than reactions elsewhere, allowing the perception of a stability that is illusory. (For a review of experimental evidence relating perception to language, note Krauss, 1968.)

In summary, "aging," too—the result of time's arrow—may then not be an objectively verifiable event but an artifact of our language. But, if the notion of aging is arbitrary, that of death and dissolution is not: Each organism or component of an organism must at some time, even if temporarily, be disassembled. This is a sine qua non for any system that has the capacity to evolve.

Consider the minimum requirements for the occurrence of evolutionary changes:

1. The organisms are self-replicating.
2. The replications are not error-free.
3. A periodic death and disassembly of the organisms into their component "parts," with these being returned to a general store.
4. A limited store of the parts—or molecules—of which the organisms are composed.

In a system with such characteristics, there will be a constant cycle of reconstruction and destruction, but, because of condition 2, some of the replicates will differ from their forebears. Most of the differences will likely be disadvantageous, just as dropping a clockwork will more often than not disturb its functions rather than improve them. Occasionally, however, one can expect a replicate to possess some advantage: requiring fewer parts, or being able to utilize components that cannot be used by others, or requiring less time for reassembly. Given that the total number of contemporaneous, coexisting organisms is limited (4), those—and their offspring—with the more "efficient" replication will account for an ever-increasing proportion of the store. When we speak of "fitness," we are comparing the proportions of some future generation of organisms that can be attributed to particular progenitors. The more "fit" of two individuals is the one whose lineage has come to monopolize the greater proportion of "parts" (4). Fitness, therefore, is a precise quantity, but it can hardly ever be measured except post facto. It certainly need not be correlated with "physical fitness" in the conventional sense. (For a further discussion, see Klopfer, 1969a and b.)

Thus, in the absence of death, and by death is meant the return of the components to the store, continuing changes in the makeup of populations would cease; there would be no evolution. But while death may be the essential precursor for evolution, why senescence or an increased probability of death with age? Evolution could still occur with at least three other types of death rates than an age-dependent mortality. The probability of dying could remain constant throughout life, just as the probability of decay of radioactive atoms remains ever constant. Or one might imagine systems in which the absolute number of deaths per unit time was constant. Finally, there are organisms, notably fish, that have inordinately high mortality rates as youngsters, but as they age—coincidently growing larger and gaining thereby immunity from an increasing array of predators—their probability for continued survival grows. The first 100 years are the hardest. In fact, for most mammals it appears that the reverse is true: The probability of dying grows with age, though it must be added that the relevant data in support of this claim are largely derived from laboratory mice and men. The technical difficulties in establishing life tables for wild animals are prodigious (Strehler, 1960).

In speculating about the cause of senescence, it cannot be overlooked that some specific, as yet unknown benefit derives from this characteristic of mortal man. But, as suggested by Medawar (1957), assume that initially a population consists of individuals whose mortality is age-independent. It will therefore consist of proportionately fewer older than younger individuals. As consequences, the latter will contribute proportionately more genetic material to the next generation's gene pool than the former. This is merely another way of asserting the greater fitness of the young relative to the old, or of saying that selection must favor the young. Given the occurrence of deleterious mutation, this will lead to the accumulation of other genes that postpone the expression of the deleterious character, even at the expense of increased dysfunction later. It also will favor genes that have pleiotropic (more than one) effects, where the earlier effect is beneficial though the later one detrimental, a genetic form of "enjoy now, pay later." This is a reasonable enough strategy where a greater proportion of the next generation's gene pool. Once such genes with their inimical though delayed effects have accumulated, of course the probability of death from disease of malfunction increases. Ergo, senescence: *quod est demonstrandum!*

In short, no less than death, senescence is an evolutionary inevitability. All the world is born to die.

A SENSE OF BEAUTY: SENSORY PHYSIOLOGY AND ESTHETICS*

A cat has settled in my lap, curled to match my contours. His head is firmly braced against my extended palm; when I cease to rub, he takes up the motion himself, certain assurance of a scratch

* Reprinted from *American Scientist,* Vol. 58, No. 4, July–August 1970, pp. 399-403, with Permission, the Society of the Sigma Xi.

behind the ears. His steady purr advertises a relaxed, somnolent state, one from which he will not readily be roused. Is it farfetched to argue that this beast is deriving pleasure from the tactile stimulation he receives? Even a hardened operationalist will concede that the animal has sought the situation that produces the behind-the-ears caress; that while thus engaged his threshold of responsiveness to other, normally attractive, lures is lessened; that no immediate function of the caress can be discerned. Of course, absence of evidence is not proof that, for instance, there are no fungi or microscopic arthropods growing behind the ears, irritants that call for a scratch. But the universality of this behavior among cats, the specificity of the body areas that when stroked provoke a purr, if not our empathy, point to an esthetic sensibility.

For ourselves, the reality of esthetics is immediate. The youngest child can be shown to prefer, if not seek out, certain patterns of peripheral stimulation, to delight in particular shapes or sounds, textures or movements. Play itself may well be no more than activity designed to produce particular proprioceptions. The questions to which I would direct us concern the generality of this phenomenon. If we consider esthetic preferences to mean a liking for objects or activities because they produce or induce particular neural inputs or emotional states, independently of overt reenforcers, can we attribute esthetics to animals other than man? The significance of an affirmative answer lies, of course, in the support that this would lend to the belief that there is a biological basis to esthetics. And should our answer be affirmative, that animals can, for instance, have "art," it will become important to inquire into the basis therefore: What are the historical or ultimate reasons for the development of an esthetic sense; by what mechanisms is the development of the species-characteristic preferences assumed?

It is essential that we develop criteria to distinguish certain classes of preferences. For instance, the octopus that is allowed to feed when a crab is presented in conjunction with a square pattern but shocked when the crab appears with a circle will learn to prefer square shapes to round (Wells, 1962). We need not refer to this as an esthetic preference as it is demonstrably dependent on an accidental (or contrived) contingency. Similarly, the cat's preference for meat over grass and the reverse desires of the antelope clearly relate to the different nutritional value of meat and grass to these two species. The lowered blood glucose and depleted muscle glycogen that come from fasting are more quickly restored by meat (for cats) or grass (for antelopes), respectively. Since we know that the input (from the food) has necessary and specified metabolic effects, it becomes unnecessary to assume esthetics to be involved.

Of course, were the cat to prefer duck to chicken, or the antelope crabgrass to fescue, assuming equal nutritional value, availability, and accessibility, we might then speak of the preference as esthetic. The implication is that esthetic preferences are those for which no immediate functional advantage can be perceived, which poses an interesting predicament. Who can deny the possibility that some significance to a preference for crabgrass may not, at a future date, be established? Perhaps fescue harbors more parasites, the equivalence in its nutritional state and the like, notwithstanding. If a negative definition of esthetics is to be avoided, a more precise notion of the importance of stimulus contingencies and reenforcements in the development of preferences is needed.

What we wish to be able to affirm is that there are patterns of stimulation, whether derived from sounds, sights, tastes, or movements, that are in themselves reenforcing: The organism, at least occasionally, seeks out these patterns, or selects them over others. This is what constitutes an esthetic preference. The octopus that has been trained to prefer squares must, at least periodically, be given a crab if his preference for squares is to persist. One complication, of course, is that even if an untrained octopus truly preferred to be near squares rather than circles, he could be trained to alter this preference, or, through training, his initial preference could be enhanced. Is it possible that success in training an animal to make certain choices demands the involvement of a particular class of neurons, association ganglia, or memory interneurons, which maintain a record of stimulus and response contingencies? Esthetic preferences, in contrast, would not be dependent on these association ganglia.

The establishment of preferences through learning, or, for that matter, learning itself, must involve the consolidation of two or more impulse patterns. A crab evokes one such pattern in a hungry octopus and the square plus feeding and the circle plus shock another.

Most hypotheses that explain learning assume the existence of interneurons that in some way establish new neural connections or selectively enhance old connections (note the excellent review of these hypotheses by Horridge, 1968). These are the "memory interneurons." Some workers (note John, 1967), building on a model formulated earlier by Lashley (1950) and Hebb (1949), propose that consolidation (memory) results from physical (e.g., electrolytic or biochemical) changes "common to a set of cells in coherent activity" (John, 1967) —changes that make it probable that this set of cells will continue to fire coherently when any of its number are activated.

Whatever the details of any process of discrimination learning, the consolidation process will represent a vital variable that presumably need not be invoked for explanation of an esthetic preference. Thus procedures that inhibit or abolish or interfere with the consolidation process in learning trials should be ineffective in unlearned choice tests. One procedure that is thought to prevent consolidation of a memory trace (i.e., to prevent learning) is the administration of electroconvulsive shock. Admittedly, there may now be some reason to question both the effectiveness and mode of action of ECS (Schneider and Sherman, 1968), but if ECS fails us, presumably some other tool will not. There is, in short, every reason to believe that we can devise acceptable operations for distinguishing preferences that result from reenforcement (and consolidation) from those that we call esthetic.

Given the prospect that we can be rigorous when we believe it needful, what are the reasons for believing in the existence of pattern or color preferences that might be esthetic? Studies with neonatal infants certainly provide compelling examples. Visual fixation times vary strikingly with the character of the pattern exposed to the infant's gaze (Salapatek, 1968). Nonhuman primates, when allowed to draw or paint, produce patterns that have a consistent form: They are not random scribbles, though, presumably, the motor abilities of these animals are not so limited as to constrain their drawings to the degree that is evident (Morris, 1962). Birds and fish have been allowed to choose between various patterns, in the absence of differential reenforcement, and appear to have consistent preferences (Rensch, 1958).

Perhaps the most convincing evidence for the existence of esthetic preferences comes from the literature on habitat selection, i.e., the choice by members of some species of particular habitats. "In the forests of New England, bird-watchers would expect to find Cape May Warblers (*Dendroica tigrina*) flitting about near the outer edge of the tops of spruce trees. Nearer the center of the trees and ranging outward from the trunk, they would hope, instead, to espy Bay-breasted Warblers (*Dendroica astanea*)" (Klopfer and Hailman,

1965). But neither warbler would be sought in a cornfield. Lack (1937), who has been concerned with the distributional effects of segregation by habitat, has detailed many "psychological preferences." By this he means situations in which a species prefers a particular food, nesting site, or some other feature of a habitat, without being dependent on or limited by it: "It. . .seems probable that each species selects its own habitat, guided by recognition features which are not in themselves essential to its existence." [P. 134.]

An analysis of habitat preferences in a North American finch (the chipping sparrow, *Spizella passerina*) revealed that features of the foliage—apparently its shape, rather than density—were among the preferenda of the birds, and this was as true for lab-reared *Kaspar Hauser* birds as for adults trapped in the wild. The *Kaspar Hauser* could be conditioned to alter their preferences, though this modification is temporary and the original bias eventually returns (Klopfer and Hailman, 1965). A similar situation exists with respect to neonatal ducklings (*Anas plalyrhynchos*) and their imprinting surrogates. While these ducklings may follow either of two different models equally enthusiastically, only one of them comes to be subsequently preferred (Klopfer, 1967, Zolman, 1969).

For gulls of the genus *Larus* there may be three sorts of habitat: one for feeding, another for courting, and yet one more for nesting. The cues used are not necessarily the same in all cases, nor are the processes by which the gulls come to know them. Some of these may also fall into the realm of esthetic preferences: e.g., the preference of the newly hatched gull chick for small, contrastingly colored moving objects (with particular shapes, colors, and rates of movement being especially preferred). The consequences of such preferences are considered below. Suffice here the conclusion that preferences do exist.

To what extent is the character of the preferences considered here imposed by morphology? This is far from a trivial question, for much of what we are able to say when we turn to questions of the functional significance of esthetics will hinge on the reply.

The newly hatched gull chick, as Hailman (1967) has shown, prefers to peck at red (650 millimicrons) and blue (450 millimicrons) models over those of other colors. This is not due to the spectral sensitivities of the birds, which are not very different from man's; i.e., the maximum sensitivity is at the region of the spectrum between 500 and 600 millimicrons. The color vision of birds, unlike that of man, however, is mediated by colored drops of oil suspended in the retinal cells. The gulls studied by Hailman had two kinds of oil droplets, red and yellow (most birds have three, and some four, kinds). Suppose that the red droplets act as light filters in retinal cells that are "excitatory" while the yellow droplets act in "inhibitory" cells. Hailman showed that both droplets freely transmit in the red and far blue part of the spectrum and that the yellow droplet also transmits orange, yellow, and some green. Thus as one moves from the blue to the red end of the spectrum, the net input from the retina should shift from being predominantly excitatory to inhibitory and back to excitatory, a hypothesis that accords well with the behavioral indication of the preferred color.

Preferences can, of course, be more complex than those for particular colors alone. Maturana and Frenk (1963) describe ganglion cells in the retina of the pigeon that show selective responsiveness to edges moving in particular directions. Some cells respond only to vertically moving edges and others only to horizontally moving ones. Hubel and Wiesel (1965) had previously found comparable movement and shape detectors in the eyes of an amphibian and a mammal. If pattern (and movement) discrimination or recognition depends on the organization of the interneurons, the effect can be similar to that where a peripheral filter in the form of an oil droplet determines color preferences. Similarly, Swihart (1967) has interpreted the electroretinograms of butterflies to mean that receptor inputs are selected to maximize responses to colors similar to those of the insects' own wings.

Finally, one can point to the organization of neurons in the brain itself as the source of preferences, though we admittedly now move farther into the realm of speculation. Sutherland (1962) has shown that an octopus can learn to discriminate between certain patterns with relative ease, as compared to others that appear to be indistinguishable to the beast. The class of the latter differs from that of the former in one interesting feature. If the paired discriminanda, two-dimensional shapes, are graphed so as to allow for a quantitative statement of their extensions in the vertical and horizontal dimensions, it is seen that the indistinguishable pairs have similar values. An upright and inverted "W" have similar vertical and horizontal extensions, for instance, though a vertical and a horizontal stripe do not. The latter are distinguishable to octopus, though the former are not. The entire visual system mirrors this organization of patterns into vertical and horizontal components, from the retinal elements to the optic lobes of the brain, whose dendritic connections are predominantly at right angles. There is even a correspondence between there being a predominance of vertical elements and octopus being able to make accurate discrimination in this plane.

The why and wherefore of the sensory filters and programmed perceptual mechanisms can easily be guessed. If there is any order in the world of which an animal is a part, it will be advantageous to the beast to be able to preceive and predict it. When "pecking at red spots" results in gaining food, design features that facilitate drawing this response will confer a selective advantage. That a generalized preference for red objects may also result is of little evolutionary consequence unless some compensatory harm results. Such pleiotropism, the secondary effect of selection for some other trait, is common enough. But it is possible for natural selection to produce more than a system for preferential detection of vital cues. If the detector circuits were to be tied to affective centers, so that their being fired itself provided a "reward" (in the sense of increasing the probability of the act that fired the detectors), the organism would not only prefer but might come actively to seek the relevant cues. In the parlance of the ethologist, it would show "appetitive behavior," a searching activity consummated not by achievement of some material goal but by a particular neural input (Craig, 1918; Thorpe, 1956). Thorpe's description of an inexperienced male dove locating its first nesting site is instructive:

> At first the dove, standing on his perch, spontaneously assumes the nest-calling attitude—body thrown forward and head down, as if neck and breast were already touching the hollow of the nest—and while in this attitude the nest-call is sounded; but all the time he shows dissatisfaction, as if the perch were not a comfortable situation for this behaviour. He shifts about and performs the action, first in one place and then in another, until he finds a corner which more or less fits his body when in this posture. He is seldom satisfied with the first corner found, but tries one after another and then perhaps comes back to the first. If now a suitable nest-box, or ready-made nest, is put into the cage, the inexperienced bird does not recognize it as a nest, but sooner or later he tries it for nest-calling, and in such trial the nest evidently gives him a strong and satisfying stimulation—a stimulation which no other circumstance has supplied. In the nest his

> attitude becomes enhanced; he turns from side to side, moving the straws with his beak, feet, breast and wings as if, to quote Wallace Craig, "rioting in the wealth of new luxurious stimuli." [Pp. 32–33.]

Nor is the suggestion of "affective centers" to be regarded as unduly speculative. The studies by Olds (1958) and his associates have revealed a multitude of loci in the brain that, on stimulation through implanted electrodes, reenforce whatever other actions occurred in temporal contiguity. Rats will bar-press for such stimulation more avidly than for food and drink. If it is needful for doves to prefer certain sites for their nest, not only should they have been designed to attend selectively to those cues that identify the correct site but, on their perceiving these cues, that act of perception should itself be promptly reenforced, so that the bird will indeed riot in the wealth of luxurious stimuli.

Is it frivolous to conclude with a note on play? While our concern with esthetics has, implicitly, been directed to the action of distance receptors—eyes, ears, nose—the example of the dove should remind us that proprioceptive input is no less relevant. If particular movements are useful to an animal, it is desirable that he recognize when these movements are correctly performed. It would be possible to design a beast that could perform only particular movements: The strike behavior of mantids is an example of such a rigidly stereotyped motor pattern (Mittelstaedt, 1962).

For animals like these, stereotypy has obviously been a good strategy. For others, particularly mammals that can less often depend on a constancy of environmental conditions, plasticity offers a higher premium. A method of accelerating the rate at which particular movements come to be coordinated can then be an advantage. Thus, instead of rigidly programming the motor circuits, one could have the proprioception from the desired patterns fed into a reenforcement center. This would assure repetition of this perception and of the acts that produced it.

Play may be the tentative explorations by which the organism "tests" different proprioceptive patterns for their goodness of fit. At least for the human athlete, there is no denying the intense, if subjective, sensation of pleasure when the fit is made. The study by Müller-Swarze (1968), who diverted his hand-raised deer with a bottle whenever they sought to play, lends some slight support to his view. His deer did not increase their play frequency, as would be expected were play to be considered the expression of a "drive state," "need," or so forth, in the manner of sleeping or dreaming, rather than a search for certain central inputs. The trauma produced by the deprivation of stimuli (the sensory isolation experiments) points as well to the importance to normal brain function of patterned inputs. Though almost a century has passed since Spencer sought to relate play to esthetics, his views wear well.

Finally, let me suggest that thought and abstraction in man is but a form of play. Perhaps certain patterns of cortical activity do produce behavior of selective value, behavior that is harmonious with the external (to us) features of the universe in which we share. Abstractions may be the play through which we learn how to think well. Perhaps esthetics, the pleasure resulting from biologically appropriate activity, will prove an important and objective guide to decision making (C. M. Bateson, in press).

Bibliography

ALLEE, W. C. (1931)
Animal Aggregations. Chicago: University of Chicago Press.

ALLEE, W. C., M. N. ALLEE, F. RITCHEY, and E. W. CASTLES (1947)
"Leadership in a flock of white Pekin Ducks," *Ecology,* **28,** 310–15.

ALTMAN, M. (1958)
"Social integration of the moose calf," *Animal Behaviour,* **6,** 155–59.

ALTMAN, M. (1960)
"The role of juvenile elk and moose in the social dynamics of their species," *Zoologica,* **45,** 35–40.

ANDREW, R. J. (1964)
"The development of adult responses from responses given during imprinting by the domestic chick," *Animal Behaviour,* **12,** 542–48.

ANDREWARTHA, H. G., and L. C. BIRCH (1954)
The Distribution and Abundance of Animals. Chicago: University of Chicago Press.

ATZ, J. (1970)
"The application of the idea of homology to behavior," in L. Aronson and E. Tobach et al., eds., *Development and Evolution of Behavior: Essays in Memory of T. S. Schneirla.* San Francisco: W. H. Freeman and Company, Publishers.

AYALA, F. J. (1968)
"Genotype, environment and population numbers," *Science,* **162,** 1453–59.

BARTHOLOMEW, C. A., and J. B. BIRDSELL (1953)
"Ecology and the proto-hominids," *American Anthropologist,* **55,** 481–98.

BATESON, C. M. (in press)
"Our own metaphor: The role of consciousness in human adaptation," *Proceedings of the Wanner-Gren Symposium, 1968.* New York: Alfred A. Knopf, Inc.

BATESON, P. P. G. (1966)
"The characteristics and context of imprinting," *Biological Reviews,* **41,** 177–220.

BEACH, I. A. (1951)
"Instinctive behavior: Reproductive activities," in S. Stevens, ed., *Handbook of Experimental Psychology,* Chap. 12, pp. 387–434. New York: John Wiley & Sons, Inc.

BEER, J., L. FRENGEL, and N. HANSEN (1956)
"Minimum space requirements of some nesting passerines," *Wilson Bulletin,* **68,** 200–209.

BENT, A. C. (1921)
"Life histories of North American gulls and terns," *Bulletin of the United States National Museum,* **113,** 1–333.

BERGMAN, G. (1955)
"Die Beziehung zwischen Bodenfarbe der Reviere und Farbe des Kuecken bei *Hydroprogne t.* und *Sterna m.," Ornis Fennica,* **32,** 69–81.

BERNSTEIN, J. S., and W. A. *Mason* (1963)
"Group formation by Rhesus monkeys," *Journal of Animal Behavior,* **11,** 28–31.

BIRDSELL, J. B. (1953)
"Some environmental and cultural factors influencing structure

of Australian aboriginal population," *American Naturalist Supplement,* **87**, 169–207.

BIRIVKOV, D. A. (1960)
Ecological Physiology of Higher Neural Activity. Leningrad: Medgiz.

BLACK-CLEWORTH, P. (1970)
"The role of electric discharges in the non-reproductive social behaviour of *Gymnotus cerapo,*" *Animal Behaviour Monographs,* **3**, 1–77.

BLAIR, W. F. (1958)
"Mating call in the speciation of anuran amphibians," *American Naturalist,* **92**, 27–51.

BLEST, A. D. (1957)
"The function of eyespot patters in the Lepidoptera," *Behaviour,* **11**, 209–55.

BLEST, A. D. (1960)
"The evolution, ontogeny, and quantitative control of the settling movements of some new saturniid moths, with some comments on distance communication by honey bees," *Behaviour,* **16**, 188–253.

BLOOMFIELD, T. M. (1968)
"Discriminatory learning in animals: an analysis through side effects," *Nature,* **217**, 929–30.

BOYCOTT, B. B. (1954)
"Learning in *Octopus vulgaris* and other cephalopods," *Stazione Zoologica Napoli,* **25**, 67–93.

BROADBENT, D. E. (1958)
Perception and Communication. Elmsford, N.Y.: Pergamon Press, Inc.

BROSHANAN, L. F. (1961)
The Sounds of Language. Cambridge, England: W. Heffer & Sons Ltd.

BROWER, J., and Z. BROWER (1960)
"Experimental studies of mimicry IV: the reactions of starlings to different proportions of models and mimics," *American Naturalist,* **94**, 271–82.

BROWER, L. P. (1958)
"Bird predation and foodplant specificity in closely related procryptic insects," *American Naturalist,* **92**, 183–87.

BROWER, L. P. (1970)
"Plant poisons in the terrestrial food chain and implications for

mimicry theory," in K. L. Chambers, ed. *Biochemical Evolution.* Corvallis: Oregon State University Press.

BROWER, L. P., L. M. COOK, and H. T. CROZE (1967)
"Predator responses to artificial Batesian mimics released in a neotropical environment," *Evolution,* **21,** 13–21.

BROWN, J. L. (1963)
"Social organization and behavior of the Mexican Jay," *Condor,* **65,** 126–53.

BROWN, J. L. (1964)
"The evolution of diversity in avian territorial systems," *Wilson Bulletin,* **76,** 160–69.

CAIN, A. J., and P. M. SHEPPARD (1954)
"Natural selection in *Cepaea,*" *Genetics,* **39,** 89–116.

CARPENTER, C. R. (1934)
"A field study of behavior and social relations of howling monkeys," *Comparative Psychology Monographs,* **19,** 1–168.

CASPARI, E. (1963)
"Selective forces in the evolution of man," *American Naturalist,* **97,** 15–28.

CHANCE, M. R. A., and W. M. S. RUSSELL (1959)
"Protean displays: a form of allesthetic behaviour," *Proceedings of the Zoological Society of London,* **132,** 65–70.

CHESLER, P. (1969)
"Maternal influences in learning by observation in kittens," *Science,* **166,** 901–3.

CHITTY, C. (1959)
"A note on stock disease," *Ecology,* **40,** 728–31.

CHRISTIAN, J. J. (1960)
"Adrenocortical and gonadal responses of female mice to increased population density," *Proceedings of the Society for Experimental Biology and Medicine,* **104,** 330–32.

CHRISTIAN, J. J. (1970)
"Social subordination, population density and mammalian evolution," *Science,* **168,** 84–94.

CLEACH, H. K. (1967)
"Temporal dissociations and population regulation in certain Hesperrine butterflies," *Ecology,* **48,** 1000–1006.

Collias, N. E. (1956)
"Analyses of socialization in sheep and goats," *Ecology,* **37,** 228–39.

Collias, E. C., and N. E. Collias (1957)
"The response of chicks of the Franklin's gull to parental bill color," *Auk,* **74,** 371–75.

Collias, N. E., and E. C. Collias (1963)
"Selective feeding by wild ducklings of different species," *Wilson Bulletin,* **75,** 6–14.

Comfort, A. (1956)
The Biology of Senescence. London: Routledge & Kegan Paul Ltd.

Commoner, B. (1969)
Burg Wartenstein Conference on the Role of Consciousness in Human Adaptation. New York: Wenner-Gren Foundation.

Commoner, B. (1970)
"Soil and fresh water: damaged global fabric," *Environment,* **12,** 4–11.

Connell, S. J., and E. Orias (1964)
"The ecological regulation of species diversity," *American Naturalist,* **99,** 399–414.

Cook, L. M., L. B. Brower, and J. Alcock (1969)
"An attempt to verify mimetic advantage in a neotropical environment," *Evolution,* **23,** 399–45.

Cooper, K. (1956)
"An instance of delayed communication in solitary wasps," *Nature,* **178,** 601–2.

Coppinger, R. P. (1970)
"The effect of experience and novelty on avian feeding behavior with reference to the warning coloration in butterflies," *American Naturalist,* **104,** 323–35.

Cott, H. B. (1956)
Adaptive Coloration in Animals. London: Methuen & Co. Ltd.

Coulson, J. C., and E. White (1959)
"The effect of age and density of breeding birds on the time of breeding of the kittiwake, *Rissa tridactyla,*" *Ibis,* **101,** 496–97.

Craig, W. (1918)
"Appetites and aversions as constituents of instincts," *Biology Bulletin,* **34,** 91–107.

Crook, J. H. (ed.) (1970)
Social Behaviour in Birds and Mammals—Essays on Social Ethology of Animals and Man. New York: Academic Press, Inc.

Crowell, K. (1961)
"The effects of reduced competition in birds," *National Academy of Sciences,* **47,** 240–43.

Croze, H. (1970)
Searching Image in Carrion Crows. Berlin: Paul Parey Verlag.

Cullen, E. (1957)
"Adaptations in the kittiwake to cliff-nesting," *Ibis,* **99,** 275–302.

Darling, F. F. 1(937)
A Herd of Red Deer. New York: Oxford University Press, Inc.

Darling, F. F. (1938)
Bird Flocks and the Breeding Cycle. New York: Cambridge University Press.

Darlington, P. J. (1957)
Zoogeography: The Geographical Distribution of Animals. New York: John Wiley & Sons, Inc.

Darlington, P. J. (1959)
"Area, climate and evolution," *Evolution,* **13,** 488–510.

Dethier, V. G. (1957)
"Communication by insects: physiology of dancing," *Science,* **125,** 331–36.

Deutsch, J. A. (1960)
The Structural Basis of Behavior. Chicago: University of Chicago Press.

Diamond, J. M. (1970)
"Ecological consequences of island colonization by S. W. Pacific birds. I. types of niche shifts," *Proceedings of the National Academy of Sciences,* **67,** 529–36.

Diamond, M. (1970)
"Intromission pattern and special vaginal code in relation to introduction of pseudo-pregnancy," *Science,* **169,** 995–97.

di Castri, F. (1967)
"Diversity of antarctic microfauna." Manuscript.

Dilger, W., and P. Johnsgard (1959)
"Comments on 'species recognition' with special reference to the Wood Duck and the Mandarin Duck," *Wilson Bulletin,* **71,** 46–53.

DUNCAN, C. J., and P. M. SHEPPARD (1965)
"Sensory discrimination and its role in the evolution of Batesian mimicry," *Behaviour,* **24,** 269–82.

EHRLICH, P. R., and P. H. RAVEN (1969)
"Differentiation of populations," *Science,* **165,** 1228–32.

EISNER, T., R. ALSOP, and G. ETTERSHANK (1964)
"Adhesiveness of spider silk," *Science,* **146,** 1058–61.

EISNER, T., T. C. KAFATOS, and E. G. LINSLEY (1962)
"Lycid predation by mimetic adult Cerambycidae (Coleoptera)," *Evolution,* **16,** 316–24.

ELIOT, T. S. (1930)
The Complete Poems and Plays. New York: Harcourt Brace Jovanovich, Inc.

ELLIS, P. (1959)
"Learning and social aggregation in locust hoppers," *Animal Behaviour,* **7,** 91–106.

ELTON, G. (1958)
The Ecology of Invasions by Animals and Plants. London: Methuen & Co. Ltd.

EMLEN, S. (1969)
"Bird migration: influence of physiological state upon celestial orientation," *Science,* **165,** 716–18.

EMLEN, S. (1971)
"The role of song in individual recognition in the Indigo," *Zeitschrift für Tierpsychologie,* **28,** 241–46.

EWER, R. H. (1956)
"Ethologic concepts," *Science,* **126,** 599–603.

FANKHAUSER, G. (1955)
"Effect of size and number of brain cells on learning in larva of *Triturus v.,*" *Science,* **122,** 692.

FEDOROV, A. (1966)
"The structure of the tropical rain forest and speciation in the humid tropics," *Journal of Ecology,* **54,** 1–12.

FELDMAN, D., and P. H. KLOPFER (in press)
"A study of observational learning in lemurs," *Zeitschrift F. Tierpsysh.* (in press).

FISCHER, A. G. (1960)
"Latitudinal variations in organic diversity," *Evolution,* **14,** 64–81.

Fisher, J. (1954)
"Evolution and bird sociality," in J. Huxley, A. C. Hardy, and E. B. Ford eds. *Evolution as a Process,* pp. 71–83. London: George Allen & Unwin Ltd.

Fisher, J., and R. A. Hinde (1950)
"The opening of milk bottles by birds," *British Birds,* **42,** 347–57.

Fisher, R. A. (1930)
The Genetical Theory of Natural Selection. New York: Oxford University Press, Inc.

Friedmann, H. (1955)
The Honeyguides. Washington, D.C.: Bulletin No. 208. U.S. National Museum.

Frings, H., M. Frings, B. Cox, and L. Peisner (1955)
"Auditory and visual mechanisms in food finding behavior of the herring gull," *Wilson Bulletin,* **67,** 155–70.

Frings, H., M. Frings, J. Jumber, R. Busnel, J. Giban, and P. Cramet (1958)
"Reactions of American and French species of *Corvus* and *Larus* to recorded communication signals tested reciprocally," *Zoology,* **39,** 126–31.

Gans, C. (1964)
"Empathetic learning and the mimicry of African snakes," *Evolution,* **18,** 705.

Garcia, J. (1971)
"Conditioning and learning factors in the regulation of food intake," *New Science* (in press).

Garcia, J., N. K. McGowan, F. R. Ervin, and R. A. Koelling (1968)
"Cues: their relative effectiveness as a function of the reenforcer," *Science,* **160,** 794–95.

Garn, S. (1963)
"Culture and the direction of human evolution," *Human Biology,* **35,** 221–36.

Gartlan, J. S. (1968)
"Structure and function in primate society," *Folia Primatalogica* **8,** 89–120.

Gause, G. J. (1934)
The Struggle for Existence. Baltimore: The Williams & Wilkins Company.

GILBERT, R. M., and N. S. SUTHERLAND (1969)
Animal Discrimination Learning. New York: Academic Press, Inc.

GINSBERG, B., E. BANKS, and D. PIMLOTT (1967)
"Symposium ecology and behavior of the wolf," *American Zoologist,* **7**, 221–381.

GOLEVA, N. G. (1955)
"Unconditioned respiratory reflexes in foxes," in *Problems of Comparative Physiology and Pathology of Higher Neural Activity,* pp. 70 ff. Leningrad: Medgiz.

GOODHART, C. B. (1958)
"Thrush predation on the snail *Cepaea hortensis,*" *Journal of Animal Ecology,* **27**, 47–57.

GOODWIN, D. W., B. POWELL, D. BROWER, H. HOINE, and J. STERN (1969)
"Alcohol and recall: state-dependent effects in man," *Science,* **163**, 1358–60.

GOTTLIEB, G. (1961a)
"The following response and imprinting in wild and domestic ducklings of the same species (*Anas platyrhynchos*)," *Behaviour,* **18**, 205–28.

GOTTLIEB, G. (1961b)
"Developmental age as a baseline for determination of the critical period in imprinting," *Journal of Comparative Physiological Psychology,* **54**, 422–27.

GOTTLIEB, G. (1970)
"Conceptions of prenatal behavior," in L. R. Aronson et al., eds., *Development and Evolution of Behavior,* pp. 111–37. San Francisco: W. H. Freeman and Company, Publishers.

GOTTLIEB, G., and P. H. KLOPFER (1962)
"The relation of developmental age to auditory and visual imprinting," *Journal of Comparative and Physiological Psychology,* **55**, 821–26.

GRANIT, R. (1955)
Receptors and Sensory Perception. New Haven, Conn.: Yale University Press.

GRANT, P. R. (1968)
"Bill size, body size and the ecological adaptation of bird species to competitive situations on islands," *Systematic Zoology,* **17**, 319–33.

HAILMAN, J. P. (1959)
"Why is the male wood duck strikingly colorful?," *American Naturalist,* **93,** 383–84.

HAILMAN, J. P. (1960)
"Ring-billed gulls following the plow," *Raven,* **31,** 109.

HAILMAN, J. P. (1961)
"Why do gull chicks peck at visually contrasting spots? A suggestion concerning social learning of food-discrimination," *American Naturalist,* **95,** 245–47.

HAILMAN, J. P. (1962)
"Pecking of laughing gull chicks to models of the parental head," *Auk,* **79,** 89–98.

HAILMAN, J. P. (1964a)
"Coding of the color preference of the gull chick," *Nature,* **204,** 710.

HAILMAN, J. P. (1964b)
"The Galapagos Swallow-tailed Gull is nocturnal," *Wilson Bulletin,* **76,** 347–54.

HAILMAN, J. P. (1966)
"Mirror-image color-preferences for background and stimulus-object in the gull chick," *Experientia,* **22,** 257.

HAILMAN, J. P. (1967)
The Ontogeny of an Instinct, Behaviour Supplement. Leiden: E. J. Brill, N.V.

HAMBURGER, V. (1963)
"Some aspects of the embryology of behavior," *Quarterly Review of Biology,* **38,** 342–65.

HAMILTON, T. (1961)
"On the functions and causes of sexual dimorphism in breeding plumage characters of North American species of warblers and orioles," *American Naturalist,* **95,** 121–23.

HAMILTON, W. D. (1964)
"The genetical evolution of social behavior I and II," *Journal of Theoretical Biology,* **7,** 1–16, 17–52.

HARDIN, G. E. (1956)
"Meaninglessness of the word protoplasm," *Science Monthly,* **82,** 112–20.

HARDIN, G. E. (1960)
"The competitive exclusion principle," *Science,* **131,** 1291–97.

HARDIN, G. E. (1968)
"The tragedy of the commons," *Science,* **162**, 1243–48.

HARLOW, H. (1959)
"Basic social capacity of primates," *Human Biology,* **31**, 40–53.

HARPER, J. (1969)
"The role of predation in vegetational diversity," *Brookhaven Symposia in Biology,* **22**, 48–62.

HASLER, A. (1956)
"Perception of pathways by fishes in migration," *Quarterly Review of Biology,* **31**, 200–209.

HAYES, K. J., and C. HAYES (1951)
"The intellectual development of a home-raised chimpanzee," *Proceedings of the American Philosophical Society,* **95**, 105–9.

HEBB, D. O. (1949)
The Organization of Behavior. New York: John Wiley & Sons, Inc.

HECHT, M. K., and D. MARIEN (1956)
"The coral snake problem: a reinterpretation," *Journal of Morphology,* **98**, 335–65.

HELMREICH, R. L. (1960)
"Regulation of reproductive rate by intra-uterine mortality in the deermouse," *Science,* **132**, 417–18.

HEMMES, R. B. (1969)
"The ontogeny of the maternal-filial bond in the domestic goat," Ph.D. thesis. Durham, N.C.: Duke University.

HENSLEY, M. M., and J. B. COPE (1951)
"Further data on removal and repopulation of the breeding birds in a spruce-fir forest community," *Auk,* **68**, 483–93.

HESPENHEIDE, H. (1966)
"The selection of comparative seed size by finches," *Wilson Bulletin,* **76**, 265–81.

HESS, E. H. (1956)
"Natural preferences of chicks and ducks for objects of different colors," *Psychology Reports,* 2, 477–83.

HESS, E. H. (1959)
"Imprinting," *Science,* **130**, 133–41.

HINDE, R. A. (1956)
"The biological significance of the territories of birds," *Ibis,* **98**, 340–69.

HINDE, R. A. (1959)
"Behaviour and speciation in birds and lower vertebrates," *Biological Reviews*, **34**. 85–128.

HINDE, R. A., and J. FISHER (1952)
"Further observations on the opening of milk bottles by tits," *British Birds*, **44**, 393–96.

HINE, B., and R. M. PAOLINO (1970)
"Retrograde amnesia: product of skeletal but not cardiac response gradient by electroconvulsive shock," *Science*, **169**, 1224–26.

HOFFMAN, H. S., J. L. SEARLE, S. TOFFEY, and F. KOZMA, JR. (1966)
"Behavioral control by an imprinted stimulus," *Journal of the Experimental Analysis of Behavior*, **9**, 177–89.

HORRIDGE, G. A. (1968)
Interneutrons. San Francisco: W. H. Freeman and Company, Publishers.

HOVANITZ, W., and V. C. S. CHANG (1963)
"Ovipositional preference tests with *Pieris*," *Journal of Research in the Lepidopteron*, **2**, 185–200.

HOWELL, T. R. (1971)
"An ecological study of the birds of the lowland pine savanna and adjacent rain forest in northeastern Nicaragua," *The Living Bird*, Tenth Annual, Cornell Laboratory of Ornithology.

HUBEL, D. H., and T. N. WIESEL (1965)
"Receptive fields and functional architecture in two non-striate visual areas of the cat," *Journal of Neurophysiology*, **28**, 229–89.

HUFFAKER, C. B., K. P. SHEA, and S. G. HERMAN (1963)
"Competitive displacement between ecological homologies," *Hilgardia*, **34**(5), 105–66.

HUTCHINSON, G. E. (1951)
"Copepodology for the ornithologist," *Ecology*, **32**, 571–77.

HUTCHINSON, G. E. (1957)
"Concluding remarks," *Cold Spring Harbor Symposia on Quantitative Biology*, **22**, 415–27.

HUTCHINSON, G. E. (1959)
"Homage to Santa Rosalia, or why are there so many different kinds of animals?," *American Naturalist*, **93**, 145–59.

HUXLEY, J., E. MAYR, and H. OSMOND (1964)
"Schizophrenia as a genetic morphism," *Nature*, **204**, 220–21.

ITANI, J. (1958)
"On the acquisition and propagation of a new food habit in the troop of the Japanese monkeys at Takasakiyami," *Primates,* **1**(2), 84–98.

JACOB, F., and J. MONOD (1961)
"Genetic regulatory mechanisms in the synthesis of proteins," *Journal of Molecular Biology,* **3**, 318–356.

JANDER, R., and I. VOLK-HEINRICHS (1970)
"Das strauch-spezifische visuelle Perceptor-System der Stabheuschrecke (*Carausius morosus*), *Zeitschrift für Vergleichende Physiologie,* **79**, 425–47.

JANZEN, D. H. (1970)
"Herbivores and the number of tree species in tropical forest," *American Naturalist,* **104**, 501–28.

JOHN, R. (1967)
Mechanisms of Memory. New York: Academic Press, Inc.

KALAT, J. W., and P. ROZIN (1970)
" 'Salience': A factor which can over-ride temporal contiguity in taste-aversion learning," *Journal of Comparative and Physiological Psychology,* **71**, 192–97.

KALKOWSKI, W. (1968)
"Social orientation by traces in the white mouse," *Folia Biologica,* **16**, 307–22.

KEAR, J. (1960)
"Food selection in certain finches with special reference to interspecific differences," Ph.D. thesis. London: Cambridge University Library.

KEITH, L. B. (1963)
Wildlife's Ten Year Cycle. Madison: University of Wisconsin Press.

KILHAM, P., P. KLOPFER, and H. OELKE (1968)
"Species identification and colour preferences in chicks," *Animal Behaviour,* **16**, 238–45.

KING, J. (1955)
"Social behavior, social organization and population dynamics in a black-tailed prairie dog town in the black hills of South Dakota," *Contributions Laboratory Vertebrate Biology,* No. 67. Ann Arbor: University of Michigan.

KLIMOVA, V. I. (1958)
"Ontogeny of reflex responses to natural food stimuli in dog and rabbits," in *Problems of the Comparative Physiology of Neural Activity,* **54**, Leningred: Izd. Iem, An-SSSR.

KLOPFER, P. H. (1956)
"Comments concerning the age at which imprinting occurs," *Wilson Bulletin,* **68**, 320–21.

KLOPFER, P. H. (1957)
"An experiment on empathic learning in ducks," *American Naturalist,* **91**, 61–63.

KLOPFER, P. H. (1959a)
"Social interactions in discrimination learning with special reference to feeding behavior in birds," *Behaviour,* **14**, 282–99.

KLOPFER, P. H. (1959b)
"An analysis of learning in young Anatidae," *Ecology,* **40**, 90–102.

KLOPFER, P. H. (1961)
"Observational learning in birds: the establishment of behavioural modes," *Behaviour* **17**, 71–80.

KLOPFER, P. H. (1963)
"Behavioral aspects of habitat selection: the role of early experience," *Wilson Bulletin,* **75**, 15–22.

KLOPFER, P. H. (1965)
"Behavioral aspects of habitat selection: a preliminary report on stereotype in foliage preferences of birds," *Wilson Bulletin,* **77**, 376–81.

KLOPFER, P. H. (1967a)
"Behavioral stereotype in birds," *Wilson Bulletin,* **79**, 290–300.

KLOPFER, P. H. (1967b)
"Is imprinting a Cheshire cat?," *Behavioral Science,* **12**, 122–29.

KLOPFER, P. H. (1967c)
"Stimulus preferences and imprinting," *Science,* **156**, 1394–96.

KLOPFER, P. H. (1968)
"Stimulus preferences and discrimination in neonatal ducklings," *Behaviour,* **32**, 309–14.

KLOPFER, P. H. (1969a)
Habitats and Territories: A Study of the Use of Space by Animals. New York: Basic Books, Inc., Publishers.

KLOPFER, P. H. (1969b)
"Instinct and chromosomes: What is an 'innate' act?," *American Naturalist,* **103,** 556–60.

KLOPFER, P. H. (1971a)
"Imprinting: Determining its perceptual basis in ducklings," *Journal of Comparative and Physiological Psychology,* **75**(3), 378–85.

KLOPFER, P. H. (1971b)
"Mother love: What turns it on?," *American Scientist,* **59,** 404–7.

KLOPFER, P. H., and G. GOTTLIEB (1962a)
"Learning ability and behavioral polymorphism within individual clutches of wild ducklings (*Anas platyrhynchos*)," *Zeitschrift für Tierpsychologie,* **19,** 183–90.

KLOPFER, P. H., and G. GOTTLIEB (1962b)
"Imprinting and behavioral polymorphism: auditory and visual imprinting in domestic ducks (*Anas platyrhynchos*) and the involvement of the critical period," *Journal of Comparative and Physiological Psychology,* **55,** 126–30.

KLOPFER, P. H., and J. P. HAILMAN (1965)
"Habitat selection in birds," in D. S. Lehrman, R. A. Hinde, and E. Shaw, eds., *Advances in the Study of Behaviour,* pp. 279–303. New York: Academic Press, Inc.

KLOPFER, P. H., and J. P. HAILMAN (1967)
An Introduction to Animal Behavior: Ethology's First Century. Englewood Cliffs, N.J.: Prentice-Hall, Inc.

KLOPFER, P. H., and M. S. KLOPFER (1968)
"Maternal 'imprinting' in goats: fostering of alien young," *Zeitschrift für Tierpsychologie,* **25,** 862–66.

KLOPFER, P. H., and R. H. MACARTHUR (1960)
"Niche size and faunal diversity," *American Naturalist,* **94,** 193–200.

KLOPFER, P. H., and R. A. MACARTHUR (1961)
"On the causes of tropical species diversity, niche overlap," *American Naturalist,* **95,** 223–26.

KLUIJVER, H. N., and L. TINBERGEN (1953)
"Territory and the regulation of density in titmice," *Archives Neerlandaises de Zoologie,* **10,** 265–89.

KNEUTGEN, J. (1970)
"Ein kranker Vogel gelangt durch ein Missverständnis an die Spitze der Rangordnung," *Zeitschrift für Tierpsychologie,* **27**, 840–41.

KOEHLER, O. (1956)
"Sprache und unbenanntes Denken," in *L'Instinct dans le comportement des animaux et des hommes,* pp. 648–75. Paris: Masson et Cie.

KOFORD, C. B. (1963)
"Rank of mothers and sons in bands of rhesus monkeys," *Science,* **141**, 356–57.

KONORSKI, J. (1948)
Conditioned Reflexes and Neuron Organization. New York: Cambridge University Press.

KOVACH, J. K. (1971)
"Ethology in the Soviet Union," Behaviour (in press).

KRAUSS, R. M. (1968)
"Language as a symbolic process in communication," *American Scientist,* **56**, 265–78.

KUO, Z. Y. (1932)
"Ontogeny of embryonic behavior in Aves," I–IV, *Journal of Experimental Zoology,* **61**, 395–430, **62**, 453–89; *Journal of Comparative Psychology,* **13**, 245–72, **14**, 109–22.

KUO, Z. Y. (1960)
"Studies on the basic factors in animal fighting," *Journal of Genetic Psychology,* **96**, 201–6, 207–16, 217–23, 225–39.

LACK, D. (1937)
"The psychological factor in bird distribution," *British Birds,* **31**, 130–36.

LACK, D. (1939)
"The behavior of the robin, I–II," *Proceedings of the Zoological Society of London,* **109**, 169–219.

LACK, D. (1954)
The Natural Regulation of Animal Numbers. New York: Oxford University Press, Inc.

LACK, D., and L. S. U. VENABLES (1939)
"The habitat distribution of British woodland birds," *Journal of Animal Ecology,* **8**, 39–71.

LANYON, W. E. (1960)
"The ontogeny of vocalizations in birds," *Animal Sounds and Communications,* Publication No. 7, 321–47. American Institute of Biological Sciences. University of Indiana, Indianapolis.

LASHLEY, K. S. (1949)
"Persistent problems in the evolution of mind," *Biological Reviews,* **24**, 28–42.

LASHLEY, K. S. (1950)
"In search of the engram," *Symposium of the Society for Experimental Biology,* **4**, 454–82.

LEHRMAN, D. S. (1956)
"On the organization of maternal behavior and the problem of instinct," in *L'Instinct dans le comportement des animaux et des hommes,* pp. 475–520. Paris: Masson et Cie.

LEMON, R. (1967)
"The response of cardinals to songs of different dialects," *Animal Behaviour,* **15**, 538–45.

LERNER, I. M. (1958)
The Genetic Basis of Selection. New York: John Wiley & Sons, Inc.

LINDROTH, C. H. (1957)
The Faunal Connections Between Europe and North America. New York: John Wiley & Sons, Inc.

LISSMAN, H. W. (1958)
"On the function and evolution of electric organs in fish," *Journal of Experimental Biology,* **35**, 156–91.

LOEHRL, H. (1959)
"Zur Frage des Zeitpunktes eines Praegung auf die Heimatregion beim Halsbandschnaepper (*Ficedula albicollis*)," *Journal für Ornithologie,* **100**, 132–40.

LOGAN, F. A. (1961)
"Specificity of discrimination learning to the original context," *Science,* **133**, 1355–56.

LORENZ, K. (1935)
"Der Kumpan in der Umwelt des Vogels," *Journal of Ornithology,* **83**, 137–214.

LORENZ, K. (1937)
"The companion in the bird's world," *Auk,* **54**, 245–73.

LORENZ, K. (1941)
"Vergleichende Bewegungsstudien an Anatinen," *Journal für Ornithologie,* **89,** 194–294.

LORENZ, K. (1970)
"Companions as factors in the bird's environment," in *Studies in Animal and Human Behavior,* Vol. I. Cambridge, Mass.: Harvard University Press.

MACARTHUR, R. H. (1955)
"Fluctuations of animal population, and a measure of community stability," *Ecology,* **36,** 353–56.

MACARTHUR, R. H. (1958)
"Population ecology of some warblers of northeastern coniferous forests," *Ecology,* **39,** 599–619.

MACARTHUR, R. H. (1961)
"Population effects of natural selection," *American Naturalist,* **95,** 195–99.

MACARTHUR, R. H. (1965)
"Patterns of species diversity," *Biological Reviews,* **40,** 510–33.

MACARTHUR, R. H. (1969)
"Patterns of communities in the tropics," in *Speciation in Tropical Environments,* R. H. Lowe-McConnell, ed. New York: Academic Press, Inc.

MACARTHUR, R. H. and J. W. MACARTHUR (1961)
"On bird species diversity," *Ecology,* **42,** 594–98.

MACARTHUR, R. H., and E. R. PIANKA (1966)
"On optimal use of a patchy environment," *American Naturalist,* **100,** 603–9.

MACARTHUR, R. H., and E. O. WILSON (1967)
The Theory of Island Biogeography. Princeton, N.J.: Princeton University Press.

MACURA, A. (1959)
"Delayed reactions in the tawny owl, *Strix aluco,*" *Folia Biologica,* **7,** 329–48.

MANNING, A. (1966)
"Pre-imaginal conditioning in *Drosophila,*" *Nature,* **216,** 338–340.

MARLER, P. (1961)
"The logical analysis of animal communication," *Journal of Theoretical Biology,* **1,** 295–317.

Marler, P. R., and W. J. Hamilton (1966)
Mechanisms of Animal Behavior. New York: John Wiley & Sons, Inc.

Marsden, M. (1968)
"Agonistic behavior of young rhesus monkeys after changes induced in social rank of their mother," *Animal Behaviour,* **16,** 38–44.

Marshall, A. J. (1954)
Bowerbirds. New York: Oxford University Press, Inc.

Maturana, H. R., and S. Frenk (1963)
"Directional movement and edge detectors in the pigeon retina," *Science,* **142,** 977–79.

Maturana, H. R., J. Y. Lettvin, W. S. McCulloch, and W. H. Pitts (1960)
"Anatomy and physiology of vision in the frog," *Journal of General Physiology,* **43,** 129–76.

Maynard-Smith, J. (1966)
"Sympatric speciation," *American Naturalist,* **100,** 637–50.

Mayr, E. (1963)
Animal Species and Evolution. Cambridge, Mass.: Belknap Press of Harvard University Press.

McCabe, T. T., and B. D. Blanchard (1950)
Three Species of Peromyscus. Cal.: Rood Associates Publication.

McDougall, W. (1905)
Physiological Psychology. London: J. M. Dent & Sons Ltd.

Meadows, P. S. (1967)
"Discrimination, previous experience and substrate selection by the amphipod *Corophium,*" *Journal of Experimental Biology,* **47,** 553–59.

Medawar, P. (1957)
The Uniqueness of the Individual. London: Methuen & Co. Ltd.

Meischner, I. (1964)
"Die motorische Lernleistungen der Vogel," *Beiträge zur Vogelkunde,* **9**. 303–74.

Miller, A. (1942)
"Habitat selection among higher vertebrates and its relation to interspecific variation," *American Naturalist,* **76,** 25–35.

MILLER, A. (1959)
"Reproductive cycles in an equatorial sparrow," *Proceedings of the National Academy of Sciences,* **45**(7), 1095–1100.

MILLER, R. E., J. BANKS, JR., and H. KUWAHARA (1966)
"The communication of affects in monkeys: cooperative reward conditioning," *Journal of Genetic Psychology,* **108**, 121–34.

MILLER, R. E., J. BANKS, JR., and N. OGAWA (1963)
"Role of facial expression in 'cooperative avoidance conditioning' in monkeys," *Journal of Abnormal and Social Psychology,* **67**, 24–30.

MILNER, P. M. (1960)
"Learning in neural systems," in M. C. Yovits and S. Cameron, eds., *Self-Organizing Systems,* pp. 190–204. Elmsford, N.Y.: Pergamon Press, Inc.

MITTELSTAEDT, H. (1962)
"Prey capture in mantids," in B. T. Sheer, ed., *Recent Advances in Invertebrate Physiology,* pp. 51–71. Eugene: University of Oregon Press.

MIYADI, D. (1959)
"On some new habits and their propagation in Japanese monkey groups," *Proceedings of the XV International Congress for Zoology,* pp. 857–60.

MORRIS, D. (1956)
"Feather postures of birds and the problem of the origin of social signs," *Behaviour,* **9**, 75–113.

MORRIS, D. (1962)
The Biology of Art. New York: Alfred A. Knopf, Inc.

MOYNIHAN, M. (1959)
"A revision of the family Laridae (Aves)," *American Museum Novitates, 1928,* 1–42.

MUELLER, H. (1968)
"Prey selection: oddity or conspicuousness," *Nature,* **217**, 92.

MÜLLER-SWARZE, D. (1968)
"Play deprivation in deer," *Behaviour,* **31**, 144–62.

NEWTON, J. (1967)
"The adaptive radiation and feeding ecology of some British Finches," *Ibis,* **109**, 33–98.

NICHOLOSON, A. J., and U. A. BAILEY (1935)
"The balance of animal populations," Part I, *Proceedings of the Zoological Society of London,* pp. 551–98.

Nissen, H. (1958)
"Basis of behavior comparison," in A. Roe and G.G. Simpson, ed., *Behavior and Evolution,* pp. 183–205. New Haven, Conn.: Yale University Press.

Noble, G. K. (1936)
"Courtship and sexual selection of the flicker, *Colaptes auratus,*" *Auk,* **53**, 269–82.

Norton-Griffiths, M. (1966)
"Some ecological aspects of the feeding behavior of the oyster catcher *Haematopus ostralegus* on the edible mussel, *Mytilus edulis,*" *Ibis,* **109**, 412–25.

Oatley, K. (1970)
"Brain mechanics and motivation," *Nature,* **225**, 797–801.

Odum, E. (1959)
Fundamentals of Ecology. Philadelphia: W. B. Saunders Company.

Olds, J. (1958)
"Self-stimulation of the brain," *Science,* **127**, 315–24.

Oppenheim, R. W. (1966)
"Amniotic contraction and embryonic motility in the chick embryo," *Science,* **152**, 528–29.

Oppenheim, R. W. (1970)
"Some aspects of embryonic behaviour in the duck (*Anas platyrhynchos*)," *Animal Behaviour,* **18**, 335–52.

Orians, G. H. (1970)
"The number of bird species in some tropical forests," *Ecology,* **50**, 782–801.

Paine, R. T. (1966)
"Food web complexity and species diversity," *American Naturalist,* **100**, 65–75.

Paine, R. T. (1969)
"The *Piaster-Tegula* interaction: prey patches, predator food preference, and intertidal community structure," *Ecology,* **50**, 950–61.

Park, T. (1948)
"Experimental studies of interspecies competition," *Ecological Monographs,* **18**, 265–308.

Patrick, R. (1964)
"A discussion of the Catherwood expedition to the Peruvian headquarters of the Amazon," *Verhalten der International Vereingung Theoretische Angewandte Limnologie,* **15**, 1084–90.

PENFIELD, W., and L. ROBERTS (1959)
Speech and Brain Mechanism. Princeton, N.J.: Princeton University Press.

PIANKA, E. (1966)
"Latitudinal gradients in species diversity: a review of concepts," *American Naturalist,* **100,** 33–46.

PIMENTEL, D. (1961)
"Animal population regulation by the genetic feedback mechanism," *American Naturalist,* **95,** 65–79.

PIROWSKI, J. (1967)
"Die Auswahl des Brutbiotops beim Feldsperling (*Passer montanus*)", *Ekologia Polska,* **A15**(1), 1–30.

PORTER, J. P. (1910)
"Intelligence and imitation in birds: a criterion of imitation," *American Journal of Psychology,* **21,** 1–71.

PRATT, C. L., and G. P. SACKETT (1967)
"Selection of social partners as a function of peer contact during rearing," *Science,* **155,** 1133–35.

PROP, N. (1960)
"Protection against birds and parasites of some species of Tenthredinid larvae," *Archives Neerlandaises de Zoologie,* **13,** 380–447.

PUBOLS, B. H. (1960)
"Incentive magnitude, learning, and performance in animals," *Psychological Bulletin,* **57,** 89–115.

PULLIAM, H. R. (1970)
"Evolution of the feeding strategy of the tropical finch *Tiaris olivacae,*" Ph.D. thesis. Durham, N.C.: Duke University.

PULLIAM, H. R., B. GILBERT, P. H. KLOPFER, D. MCDONALD, L. MCDONALD, and G. MILLIKAN (1972)
"On the evolution of sociality, with special reference to the grassquit (*Tiaris olivacea*)." *Wilson Bull.* (in press)

RECHER, H. F. (1969)
"Bird species diversity and habitat diversity in Australia and North America," *American Naturalist,* **103,** 75–80.

RENSCH, B. (1958)
"Die Wirksamkeit ästhetischer Faktoren bei Wirbeltieren," *Zeisschrift für Tierpsychologie,* **15,** 447–61.

RENSCH, B. (1959)
"Trends toward progress of brains and sense organs," *Cold Spring Harbor Symposia on Quantitative Biology,* **24,** 291–303.

RICHARDS, C. M. (1958)
"Inhibition of growth in crowded *Rana pipiens* tadpoles," *Physiological Zoology,* **31,** 138–50.

RICKLEFS, E. (1966)
"The temporal component of diversity among species of birds," *Evolution,* **20,** 235–42.

RIOPELLE, A. J. (1960)
"Observational learning of a position habit by monkeys," *Journal of Comparative and Physiological Psychology,* **53,** 426–28.

ROEDER, K. D. (1959)
"A physiological approach to the relation between prey and predator," *Smithsonian Miscellaneous Collection,* **137,** 287–306.

ROTHSCHILD, H., and B. FORD (1968)
"Warning signals from a starling *Sturnus vulgaris*: Observing a bird rejecting impalatable prey," *Ibis,* **110,** 104–5.

DE RUITER, L. (1952)
"Some experiments on the camouflage of stick caterpillars," *Behaviour,* **4,** 222–32.

SACKETT, G. P. (1966)
"Monkeys reared in isolation with pictures as visual input: Evidence for an innate releasing mechanism," *Science,* **154,** 1468–71.

SALAPATEK, P. (1968)
"Visual scanning by the human newborn," *Journal of Comparative and Physiological Psychology,* **66,** 247–58.

SANDERS, H. L. (1969)
"Benthic marine diversity and the stability-time hypothesis," *Brookhaven Symposia in Biology,* **22,** 71–81.

SARGENT, T. D. (1968)
"Cryptic moths: Effects on background selections of painting the circumocular scales," *Science,* **159,** 100–101.

SATTLER, K. M. (1970)
"Olfactory and auditory stress on mice," Thesis. Durham, N.C.: Duke University.

SCHACTER, S. (1968)
"Obesity and eating," *Science,* **161,** 751–55.

SCHEIN, M. W. (1963)
"On the irreversibility of imprinting," *Zeitschrift für Tierpsychologie,* **20,** 462–67.

SCHNEIDER, A. W., and W. SHERMAN (1968)
"Amnesia: a function of the temporal relation of footshock to ECS," *Science,* **159,** 219–21.

SCHNEIDER, G. E. (1969)
"Two visual systems," *Science,* **163,** 894–902.

SCHOENER, T. W. (1965)
"The evolution of bill size differences among sympatric congeneric species of birds," *Evolution,* **19,** 189–213.

SCHOENER, T. W. (1968)
"The sizes of feeding territories among birds," *Ecology,* **49,** 123–41.

SCHROEDINGER, E. (1951)
What is Life? New York: Cambridge University Press.

SCHUTZ, F. (1965)
"Sexuelle Prägung bei Anatiden," *Zeitschrift für Tierpsychologie,* **22,** 50–103.

SEBEOK, T. (1968)
Animal Communication. Bloomington: Indiana University Press.

SEIGER, M. B. (1967)
"A computer simulation study of the influence of imprinting on population structure," *American Naturalist,* **101,** 47–58.

SELYE, H. (1956)
The Stress of Life. New York: McGraw-Hill Book Company.

SEXTON, O. (1960)
"Experimental studies of artificial Batesian mimics," *Behaviour,* **15,** 244–52.

SEXTON, O. J., H. HEATWOLE, and D. KNIGHT (1964)
"Correlation of microdistribution of some Panama reptiles and amphibians with structural organization of the habitat," *Caribbean Journal of Science,* **4,** 261–95.

SEXTON, O., E. ORTLEB, and C. HAGER (1966)
"*Anolis cardinensis:* Effects of feeding in reaction to aposematic prey," *Science,* **153,** 1140.

SHEPPARD, P. M. (1969)
"The evolution of mimicry: a problem in ecology and genetics," *Cold Spring Harbor Symposia on Quantitative Biology,* **24,** 131–40.

SHEPPARD, D. H., P. H. KLOPFER, and H. OELKE (1968)
"Habitat selection: differences in stereotypy between insular and continental birds," *Wilson Bulletin,* **80,** 452–57.

SIEGEL, H. S., and P. B. SIEGEL (1961)
"The relationship of social competition with endocrine weights and activity in male chickens," *Animal Behaviour,* **9,** 151–58.

SIMMS, E. (1955)
"Conversational calls of birds as revealed by new methods of field recording," *Acta XI Congress International Ornithology,* Basle, Switzerland, 1954. pp. 623–26.

SIMPSON, G. G. (1953)
"The Baldwin Effect," *Evolution,* **7,** 1·10–17.

SKUTCH, A. F. (1954)
"Life histories of Central American birds," *Pacific Coast Avifauna,* **31.**

SKUTCH, A. F .(1960)
"Life histories of Central American birds," *Pacific Coast Avifauna,* **34.**

SKUTCH, A. F. (1966)
"A breeding bird census and nesting success in Central America," *Ibis,* **108,** 1–16.

SKUTCH, A. F. (1967)
"Adaptive limitation of the reproductive ratio of birds," *Ibis,* **109,** 579–99.

SLOBODKIN, L. B., and H. L. SANDERS (1969)
"On the contribution of environmental predictability to species diversity," *Brookhaven Symposia in Biology,* **22,** 82–95.

SLONIM, A. D. (1961)
Foundations of General Ecological Physiology of Mammals. Moscow-Leningrad: An-SSSR.

SLONIM, A. D. (1967)
Instinct: Puzzle of Inherited Behavior. Leningrad: Navka.

SLUCKIN, W. (1965)
Imprinting and Early Learning. Chicago: Aldine Publishing Company.

SMITH, N. (1966)
"Evolution of some arctic gull (*Larus*): an experimental study of isolating mechanisms," *AOU Ornithological Monographs,* **7,** 1–99.

SMITH, W. (1957)
"Social learning in domestic ducks," *Behaviour,* **11**, 40–55.

SNYDER, R. G. (1961)
"The sex ratio of offspring of flyers of high performance military aircraft," *Human Biology,* **33**, 1–10.

SOULÉ, M., and B. R. STEWART (1970)
"The 'niche variation' hypothesis: A test and alternatives," *American Naturalist,* **104**, 85–97.

SPENCER, H. (1898)
The Princples of Psychology. New York: D. Appleton.

SPURWAY, H. (1955)
"The causes of domestication: An attempt to integrate some ideas of Konrad Lorenz with evolution theory," *Journal of Genetics,* **53**, 325–62.

STEIN, R. E. (1963)
"Isolating mechanisms between populations of Traill's fly catchers," *Proceedings of the American Philosophical Society,* **107**, 21–50.

STERN, C. (1958)
"Selection for subthreshold differences and the origin of pseudoexogenous adaptations," *American Naturalist,* **92**, 313–16.

STREHLER, B. L. (ed.) (1960).
The Biology of Aging. Washington, D.C.: American Institute of Biological Sciences.

STRONG, R. M. (1914)
"On the habits and behaviour of the herring gull, *Larus argentatus Pont.,*" *Auk,* **31**, 22–49, 178–99.

STRUHSAKER, T. T. (1969)
"Correlation of ecology and social organization among African cercopithecines," *Folia Primatologica,* **11**, 80–118.

SUTHERLAND, N. S. (1957)
"Visual discrimination of orientation and shape by the octopus," *Nature,* **179**, 11–13.

SUTHERLAND, N. S. (1958)
"Visual discrimination of shape by *Octopus*: Squares and triangles," *Quarterly Journal of Experimental Psychology,* **10**, 40–47.

SUTHERLAND, N. S. (1960a)
"Visual discrimination of orientation by *Octopus*: Mirror images," *British Journal of Psychology,* **51**, 9–18.

SUTHERLAND, N. S. (1960b)
"Visual discrimination of shape by *Octopus*: Open and closed forms," *Journal of Comparative and Physiological Psychology,* **53,** 104–112.

SUTHERLAND, N. S. (1962)
Shape Discrimination by Animals," Experimental Psychological Society Monograph 1. Cambridge, England: W. Heffer & Sons, Ltd.

SWIHART, S. (1967)
"Neural adaptation in the visual pathway of certain Heliconicine butterflies, and related forms, to variations in wing coloration," *Zoologica,* **52,** 1–14.

TERBORGH, J., and J. S. WESKE (1970)
"Colonization of secondary habitats by Peruvian birds," *Ecology,* **50,** 765–82.

THOMPSON, W. R., and W. HERON (1954)
"The effects of early restriction on activity in dogs," *Journal of Comparative and Physiological Psychology,* **41,** 77–82.

THORPE, W. H. (1956)
Learning and Instinct in Animals. London: Methuen & Co. Ltd.

THORPE, W. H. (1958)
"The learning of song patterns by birds with especial reference to the song of the chaffinsh, *Fringilla coelobs,"* *Ibis,* **100,** 535–70.

THORPE, W. H., and F. G. W. JONES (1937)
"Olfactory conditioning and its relation to the problem of host selection," *Proceedings of the Royal Society (London),* **B124,** 56–81.

THORPE, W. H. and B. I. LADE (1961)
"The song of some families of the passeriformes," *Ibis,* **103***a,* 231–59.

TINBERGEN, L. (1960)
"The natural control of insects in pine woods," *Archives Neerlandaises de Zoologie,* **13,** 265–379.

TINBERGEN, N. (1951)
The Study of Instinct. New York: Oxford University Press, Inc.

TINBERGEN, N. (1953)
The Herring Gull's World. New York: William Collins Sons & Co., Ltd.

TINBERGEN, N., and A. C. PERDECK (1950)
"On the stimulus situation releasing the begging response in the newly hatched herring gull chick (*Larus argentatus*)," *Behaviour,* **3,** 1–39.

TODD, J. H., J. ATEMA, and J. BARDACH (1967)
"Chemical communication in social behavior of a fish, the yellow bullhead," *Science,* **158,** 672–73.

TOLMAN, C. W., and G. F. WILSON (1965)
"Social feeding in domestic chicks," *Animal Behaviour,* **13,** 134–42.

TREAT, A. E. (1955)
"The reaction time of noctuid moths to ultrasonic stimulation," *Journal of the N.Y. Entomological Society,* **64,** 165–71.

TURNER, E. R. A. (1964)
"Social feedings in birds," *Behaviour,* **24,** 1–46.

VON UEXKUELL, J. (1921)
Umwelt und Innenwelt der Tiere. Berlin: Springer-Verlag.

UTIDA, S. (1957)
"Cyclic fluctuations of population density intrinsic to the host-parasite system," *Ecology,* **39,** 442–49.

VAN VALEN, L. (1965)
"Morphological variation with width of ecological niche," *American Naturalist,* **99,** 377–90.

VINCE, M. A. (1960)
"Developmental changes in responsiveness in the great tit, *Parus major,*" *Behaviour,* **15,** 219–43.

VINE, I. (1971)
"Risk of visual detection and pursuit by a predator and the selective advantage of flocking behavior," *Journal of Theoretical Biology,* **30,** 405–22.

VOLTERRA, V. (1931)
Lecon sur la théorie mathématique de la lutte pour la vie. Paris: Gauthier-Villars.

VORONIN, L. G. (1957)
"Comparative physiology of higher neural activity," *Lectures.* Moscow: Izd. MGU.

VOWLES, D. (1961)
"Neural mechanisms in insect behaviour," in W. Thorpe and O.L. Zangwill, eds., *Current Problems in Animal Behaviour,* pp. 5–29. New York: Cambridge University Press.

VUILLEUMIER, F. in press
"Avian species diversity in temperate South America: is it different from North America?," *American Naturalist* (in press).

WADDINGTON, C. H. (1959)
"Evolutionary systems: animal and human," *Nature,* **183,** 1634–38.

WADDINGTON, C. H. (1960)
The Ethical Animal. London: George Allen & Unwin Ltd.

WADDINGTON, C. H. (1966)
Principles of Development and Differentiation. New York: The Macmillan Company.

WALLS, G. L. (1942)
The Vertebrate Eye and Its Adaptive Radiation, Bulletin No. 19. Cranbrook Institute of Science.

WATSON, A. (1967)
"Population control by territorial behavior in Red Grouse," *Nature,* **215,** 1274–75.

WATTS, A. (1966)
The Book. New York: Pantheon Books, Inc.

WECKER, S. (1969)
"Habitat selection in *Peromyscus*: differential learning in two age classes of the prairie deermouse." Manuscript.

WEIDMANN, U. (1959)
"The begging response of the black-headed gull chick," paper read before the 6th International Ethological Congress. Mimeo.

WEIDMAN, U. (1961)
"The stimuli eliciting begging in gulls and terns," *Animal Behaviour,* **9,** 115–16.

WEIDMANN, R., and U. WEIDMANN (1958)
"An analysis of the stimulus situation releasing food-begging in the black-headed gull," *Animal Behaviour,* **6,** 114.

WELCH, B. L., and P. H. KLOPFER (1961)
"Endocrine variability as a factor in the regulation of population genetics," *American Naturalist,* **95,** 256–60.

WELLER, M. W. (1959)
"Parasitic egg laying in the redhead (*Aythya americana*) and other North American anatidae," *Ecological Monographs,* **29,** 333–65.

WELLS, M. J. (1959)
"A touch learning center in *Octopus*," *Journal of Experimental Biology,* **36,** 590–612.

WELLS, M. J. (1962)
Brain and Behaviour in Cephalopods. London: William Heinemann Led.

WHITTAKER, R. H. (1956)
"Vegetation of the Great Smoky Mountains," *Ecological Monographs,* **26,** 1–86.

WHITTAKER, R. H. (1960)
"Vegetation of the Siskiyou Mountains, Oregon and California," *Ecological Monographs,* **30,** 279–338.

WHORF, B. L. (1956)
Language, Thought and Reality. New York: John Wiley & Sons, Inc.

WICKLER, W. (1968)
Mimicry in Plants and Animals. New York: McGraw Hill Book Company.

WIENS, J. A. (1966)
"On group selection and Wynne-Edwards hypothesis," *American Science,* **54,** 273–87.

WIENS, J. A. (1969)
"An approach to the study of ecological relationships among grassland birds," *AOU Ornithological Monographs,* **8,** 1–93.

WOLDA, H. (1963)
"Natural populations of the polymorph landsnail *Cepaea nemoralis,* Factors affecting their size and their genetic constitution," *Archives Néerlandaises de Zoologie,* **15,** 381–471.

WOLFSON, A. (1960)
"Role of light and darkness in the regulation of the annual stimulus for spring migration and reproductive cycles," *Proceedings of the XII International Ornithological Congress, Helsinki, 1958,* pp. 758–89.

WYNNE-EDWARDS, V. C. (1962)
Animal Dispersion in Relation to Social Behavior, p. 653. New York: Hafner Publishing Company, Inc.

YOUNG, J. Z. (1961)
"Learning and discrimination in the Octopus," *Biological Reviews,* **36,** 32–96.

ZIGLER, E. V. (1967)
"Familial mental retardation: a continuing dilemma," *Science,* **155,** 292–98.

ZIPF, G. K. (1949)
Human Behavior and the Principle of Least Effort. Reading, Mass.: Addison-Wesley Publishing Company, Inc.

ZOLMAN, J. F. (1969)
"Stimulus preferences and form discrimination in young chicks," *Psychological Record,* **19,** 407–16.

Author Index

C

D

E

F

G

H

T

U

V

W

Y

Z

Subject Index

T

U

V